SECONDARY EDUCATION IN A CHANGING WORLD

Series editors: Barry M. Franklin and Gary McCulloch

Published by Palgrave Macmillan:

The Comprehensive Public High School: Historical Perspectives
By Geoffrey Sherington and Craig Campbell
(2006)

Cyril Norwood and the Ideal of Secondary Education
By Gary McCulloch
(2007)

The Death of the Comprehensive High School?: Historical, Contemporary, and Comparative Perspectives
Edited by Barry M. Franklin and Gary McCulloch
(2007)

The Emergence of Holocaust Education in American Schools
By Thomas D. Fallace
(2008)

The Standardization of American Schooling: Linking Secondary and Higher Education, 1870–1910
By Marc A. VanOverbeke
(2008)

Education and Social Integration: Comprehensive Schooling in Europe
By Susanne Wiborg
(2009)

Reforming New Zealand Secondary Education: The Picot Report and the Road to Radical Reform
By Roger Openshaw
(2009)

Inciting Change in Secondary English Language Programs: The Case of Cherry High School
By Marilee Coles-Ritchie
(2009)

Curriculum, Community, and Urban School Reform
By Barry M. Franklin
(2010)

Girls' Secondary Education in the Western World: From the 18th to the 20th Century
Edited by James C. Albisetti, Joyce Goodman, and Rebecca Rogers
(2010)

Race-Class Relations and Integration in Secondary Education: The Case of Miller High
By Caroline Eick
(2010)

Teaching Harry Potter: The Power of Imagination in Multicultural Classrooms
By Catherine L. Belcher and Becky Herr Stephenson
(2011)

The Invention of the Secondary Curriculum
By John White
(2011)

Secondary STEM Educational Reform
Edited by Carla C. Johnson
(2011)

Secondary STEM Educational Reform

Edited by

Carla C. Johnson

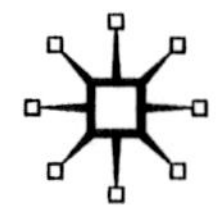

First published in 2011 by
PALGRAVE MACMILLAN®
in the United States—a division of St. Martin's Press LLC,
175 Fifth Avenue, New York, NY 10010.

Where this book is distributed in the UK, Europe and the rest of the world, this is by Palgrave Macmillan, a division of Macmillan Publishers Limited, registered in England, company number 785998, of Houndmills, Basingstoke, Hampshire RG21 6XS.

Palgrave Macmillan is the global academic imprint of the above companies and has companies and representatives throughout the world.

Palgrave® and Macmillan® are registered trademarks in the United States, the United Kingdom, Europe and other countries.

ISBN: 978–0–230–11185–1

Library of Congress Cataloging-in-Publication Data

Secondary STEM educational reform / edited by Carla C. Johnson.
p. cm.—(Secondary education in a changing world)
ISBN 978–0–230–11185–1
1. Science—Study and teaching (Secondary)—United States.
2. Technology—Study and teaching (Secondary)—United States.
3. Engineering—Study and teaching (Secondary)—United States.
4. Mathematics—Study and teaching (Secondary)—United States.
5. Educational change—United States. I. Johnson, Carla C., 1969–

Q183.3.A1S415 2011
373.12'07—dc23 2011020513

A catalogue record of the book is available from the British Library.

Design by Newgen Imaging Systems (P) Ltd., Chennai, India.

First edition: December 2011

10 9 8 7 6 5 4 3 2 1

Printed in the United States of America.

I would like to dedicate this book to three dear mentors who have impacted my life and my career significantly. First, to Dr. Piyush Swami from the University of Cincinnati who served as my dissertation chair and treasured colleague. I hope you enjoy new adventures in your retirement. Second, to Dr. Jane Butler Kahle, from Miami University, one of the pioneers of science education who agreed to mentor me during my doctoral program and has continued to be a cherished colleague. Last, but not least, to Dr. Charlene M. Czerniak, my colleague who brought me in to the University of Toledo and taught me many of the keys to being a successful and productive faculty member. I am proud to call all three of these great science educators dear friends, and I hope to build upon the important contributions they have made to our field.

Contents

Figures and Tables

Figures

Tables

Series Editors' Foreword

Among the educational issues affecting policy makers, public officials, and citizens in modern, democratic, and industrial societies, none has been more contentious than the role of secondary schooling. In establishing the Secondary Education in a Changing Worlds series with Palgrave Macmillan, our intent is to provide a venue for scholars in different national settings to explore critical and controversial issues surrounding secondary education. We envision our series as a place for the airing and resolution of these controversial issues.

More than a century has elapsed since Emile Durkheim argued the importance of studying secondary education as a unity, rather than in relation to the wide range of subjects and the division of pedagogical labor of which it was composed. Only thus, he insisted, would it be possible to have the ends and aims of secondary education constantly in view. The failure to do so accounted for a great deal of difficulty with which secondary education was faced. First, it meant that secondary education was "intellectually disorientated," between "a past which is dying and a future which is still undecided," and as a result "lacks the vigor and vitality which it once possessed" (Durkheim, 1938/1977, p. 8). Second, the institutions of secondary education were not understood adequately in relations to their past, which was "the soil which nourished them and gave them their present meaning, and apart from which they cannot be examined without a great deal of impoverishment and distortion" (10). And third, it was difficult for secondary school teachers, who were responsible for putting policy reforms into practice, to understand the nature of the problems and issues that prompted them.

In the early years of the twenty-first century, Durkheim's strictures still have resonance. The intellectual disorientation of secondary education is more evident than ever as it is caught up in successive waves of policy changes. The connections between the present and the past have become increasingly hard to trace and untangle. Moreover, the distance between policy makers on the one hand and the practitioners on the other has rarely seemed as immense as it is today. The key mission of the current series of book is, in the spirit of Durkheim, to address these underlying dilemmas of secondary education and to play a part in resolving them.

In *Secondary STEM Educational Reform,* Carla C. Johnson has brought together a collection of essays authored by a diverse group of science educators who have been involved in implementing programs at various sites throughout the United States in STEM (Science, Technology, Engineering, and Mathematics) education. At the heart of the book and central to the argument of its seven chapters is the concept of *turbulence* or the interaction between STEM reform policies and the stakeholders who develop them.

The book's contents consider two broad categories of turbulence, those found at the micro level of the school and the school district and those found at the macro level of the state and the nation. The essays explore a number of different forms of turbulence at these levels. At the micro level, they consist of such factors as the personnel policies that govern teachers and staff, salaries, scheduling, and student support policies. At the macro level, the impetus for turbulence is to be found in policies related to accountability and funding with the most important being those involved in the implementation of the No Child Left Behind legislation. The book's chapters provide an overview of how instances of turbulence manifest themselves in STEM reform projects, the success of programs that operate in the midst of different forms of turbulence, and the challenges that such projects face.

The volume includes chapters that address a number of key issues involved in STEM reform and how an understanding of the dilemmas posed by the presence of turbulence affect those initiatives. Included in the volume are essays exploring an array of STEM initiatives including programs for minority students, efforts within economically distressed communities, rural reform programs, the interplay between literacy and programs in mathematics and science, the governance of preservice science and mathematics teacher preparation programs, and the interplay between STEM programs and regional improvements in economic development and quality of life.

In the way of conclusion, Johnson offers a postscript in which she returns to the preceding chapters and suggests how each of them offers solutions to problems of turbulence. The remedies are wide ranging and include among others efforts to enhance professional development programs, to establish community-business partnerships, to offer teachers increased time for planning, and to introduce and support programs for individualized instruction.

Secondary STEM Educational Reform is the latest book in the series that focuses its attention on pedagogical and curricular issues involving secondary education reform. We believe that that there is a critical connection among issues of curriculum, pedagogy, and educational policy. *Secondary STEM Educational Reform* is a good example of a study that explores that linkage. We believe that the academic disciplines are playing a critical role in the globalized trajectory that contemporary secondary educational reform is taking. Taking our cues from the work of Durkheim, it is our intention to continue to publish books, particularly comparative studies, which connect secondary school reform with disciplinary knowledge in its broadest, most integrated forms.

BARRY M. FRANKLIN
GARY MCCULLOCH
SERIES COEDITORS

Preface: Defining Turbulence in STEM Educational Reform

Carla C. Johnson
University of Cincinnati

The Secondary STEM Educational Reform book is a collection of chapters from authors who have been engaged in implementation of programs designed to transform STEM education K-12, as well as provide suggestions on navigating inevitable *turbulence* associated with reform. Turbulence is the term we have used to describe the interplay of external variables that directly influence school reform (Johnson and Marx, 2009; Johnson, 2010; Johnson and Fargo, 2010). Specifically, turbulence is the interaction between policy (formal and/or informal) and stakeholders engaged in reform. Most policy decisions are based upon either budgetary or accountability rationales and is out of the control of teachers, as well as directors of reform programs who are often not employed by the participating school district. Turbulence has both micro (school and district) and macro levels (state and national) and the interplay of components within and across levels can facilitate or impede educational change.

Micro-level policies that may create turbulence are primarily focused in areas relating to personnel, scheduling, student support, learning environment, accountability, and community engagement. Personnel-level policies that are attributed to turbulence include teacher workload, salary, job assignment, and opportunities for professional growth. As funding for education continues to dwindle, districts and schools seek ways to better structure their funding for staff positions. Unfortunately, this results in many secondary teachers who teach multiple various courses—often outside their area of expertise—requiring more preparation to teach effectively concurrently when teacher planning time within the school day is also being cut to further maximize instructional time. Teacher salaries remain low, forcing many to secure second jobs at the end of the school day and during summer breaks to make ends meet, reducing the time for teachers to plan, collaborate, or engage in professional growth. District development funding has in many cases been redirected to support operational budgets, thus district and school sponsored workshops for teachers are declining.

Scheduling policy includes the format of the school day including opportunities for extended class time and equitable allocation of instructional time for all subjects, as well as use of staff meeting time within the contracted school day and teacher planning and collaboration time. In lower-performing schools, more time is being devoted to reading and mathematics due to national accountability policy—reducing the time for science, social studies, physical education, and the arts (Ruby, 2006; Johnson, 2007). Staff meeting time is often utilized ineffectively as an opportunity to deliver information that could have been communicated by email. Time for teachers to meet, collaborate, and plan is often nonexistent both during the school day and within designated staff meeting times.

Student support policies that can create turbulence include class-size limitations and support for enrichment, remediation, and support for English as second language and special education students. Demands to cover curriculum and teach to the test have rendered teachers virtually unable to harness the "teachable moment" in classrooms. Opportunities for enrichment and remediation are infrequent. Pull-out programs for nonnative English-speaking students are declining, as some districts have gone to an all ESL-endorsed teacher policy—meaning that all teachers must secure ESL endorsement and make accommodations within their classes for students.

Resources for learning and facilities are the key components of learning environment policy. Districts often inequitably disperse funding for curriculum, materials, technology, and other resources across their schools—further widening the accessibility to high-quality instruction for many students. Facilities vary within and across districts depending upon property tax revenue as well.

Though most accountability policies are handed down from a macro level (state or national), there are a growing number of districts who have enacted their own accountability policy including district-developed unit assessments and pacing guides. In fact, some districts have testing for each unit of each subject in each grade level. Many administrators believe the way to better scores on assessments is to practice across the school year (Anyon, 2001; Kozol, 2005). Valuable instructional time is lost and though the intent of these tests may have been to take a formative look at student learning and remediate, there simply are not enough hours in the day to do this in most schools.

The last component of micro-level turbulence is community/parental engagement. It is important to note that most micro-level systems (schools and districts) do not have formal policies in this area, research has shown the lacking involvement and partnership of those groups can be attributed in part to the inability of reform efforts to be fully implemented and sustained long term. Competing priorities and goals of stakeholders engaged in schools, lack of awareness and advocacy for reform, as well as lack of involvement of stakeholders can create turbulence.

Macro-level policy that is associated with turbulence is focused primarily on accountability and funding. The most prominent accountability policy for schools in the United States stems from the No Child Left Behind (NCLB) legislation which requires schools to make adequate yearly progress (AYP) in reading and mathematics. Science is assessed but is not included in calculating AYP for schools though reflected in many state report cards. The high-stakes focus on reading and mathematics has resulted in increased focus by districts. Funding for schools through federal programs is secondary to property tax–base funding. However, schools and districts that receive federal investments also are required to meet certain criteria within the implementation of programs through those funds. Often the demands of meeting criteria can be counterproductive to reform.

This book will provide STEM stakeholders, policy makers, and K-12 schools with an overview of turbulence experienced within STEM reform programs, as well as the challenges of large-scale reform projects in secondary schools. The success of programs operating within environments of turbulence will also be discussed. This book has been organized to begin with small-scale projects (i.e., one school), continuing across the book to include large-scale projects in multiple districts, as well as regional reform movements.

Chapter one is a case study of creating a STEM school within a large urban district with predominantly African American students. The project directors experienced turbulence in the form of historical policy decisions (e.g., technology, curriculum, teacher time out of classroom) as well as lack of community involvement and adequate facilities for innovation.

Chapter two examines a mathematics professional development program in three schools and associated challenges relating to program evaluation, as well as participant and administrator buy in.

Chapter three presents experience of program personnel who were involved in a large-scale implementation of a rural professional development program in science to over 1,200 teachers from 36 school districts in 2 states. The associated turbulence with scale-up will be discussed relating to issues with leadership, budget, and competing priorities and goals.

Chapter four introduces the reader to a science and mathematics enrichment program for students from poverty in urban school districts. Logistical challenges, as well as issues with participants and family involvement will be shared.

Chapter five takes a look at the turbulence associated with the integration of literacy within the STEM disciplines including the struggles to break down silos and changing teacher-preparation programs to include a stronger emphasis on literacy at the secondary level.

Chapter six details the challenges of the Executive Board of a preservice science and mathematics teacher preparation program and associated turbulence relating to competing goals and perceived roles of members of the board.

Chapter seven details the turbulence associated with a regional STEM movement targeting improving quality of life and economic vitality in northwest Ohio. Issues identified include: accountability for learning, awareness and advocacy for STEM, need to improve instruction and strengthen partnerships, as well as recruitment, placement, and retention of STEM talent.

Collectively, all of the chapters comprise a valuable resource for those engaged in secondary STEM education reform and further our understanding of the turbulence associated with reform. The postscript will bring together the lessons learned from each of our programs and associated chapters and will provide an overview of solutions to challenges.

CARLA C. JOHNSON
Editor
April, 2011

Acknowledgments

Secondary STEM Educational Reform represents the work of devoted stakeholders engaged in the arduous task of reforming STEM education in the United States. I would like to recognize the authors of the chapters included in this book for their persistence and dedication to improving STEM for the future generation of STEM talent. In addition to spending many hours outside of their regular faculty appointments, the authors devoted extensive time to bringing this book together, and I am greatly appreciative of their effort. I would like to express sincere gratitude to Dr. Barry M. Franklin, one of the editors on this series who has been a mentor of mine since my appointment at Utah State University. Finally, words of praise to the most important stakeholders in our STEM reform work, the teachers. Over the years, we have had the opportunity to partner with extremely inspiring and talented teachers across the United States, and this book is dedicated to those who choose to make a difference everyday through teaching.

Chapter 1

Creating a "STEM for All" Environment

James D. Basham, Catherine M. Koehler, and Maya Israel

In the fall of 2007, a growing number of groups throughout Ohio began to congregate around the idea of building a network of demonstration schools and projects that could influence college and career readiness for Science, Technology, Engineering, and Mathematics (STEM). This push was the result of both state and national initiatives in STEM education and school reform. At the national level was a growing concern over the diminished capacity of the United States to compete in the STEM global marketplace for educated citizenry and people choosing STEM careers (Business Roundtable, 2005; Committee on Science, Engineering and Public Policy [CSEPP], 2005; Dede et al., 2005). Within the state, there were multiple reports (e.g., CSEPP, 2005; Deloitte Consulting, 2005; SRI International, 2009) that indicated that the increased need for a STEM workforce was associated with the future well-being of the state.

Within our region of the state, a group formed that included public schools, the University of Cincinnati (UC), a number of corporate and industry organizations, as well as key social foundations. This initial partnership was focused on supporting STEM programs within local public schools. Specifically, efforts were targeted toward developing a prekindergarten through college pipeline for all students to be successful in STEM knowledge and general twenty-first-century skills (see Partnership for 21st Century Skills, 2004). The group's initial emphasis was placed on preparing students for college or career readiness through targeting the development of two STEM schools, one high school and one prekindergarten through eighth-grade elementary school.

Simultaneously, as the local STEM partnership was forming, the local Board of Education was also moving forward with a controversial decision to close a local elementary school, Lincoln Elementary (pseudonym). After years of academic failure, Lincoln was designated as a "school of redesign" under the tenets of the No Child Left Behind Act of 2001 (NCLB) (NCLB, 2001). In fact, this high-needs urban school was considered to be one of the lowest-performing schools in the state serving students from prekindergarten through eighth grade. Whereas some residents of the neighborhood objected to the school's closure, the decision to close the school was both an ethical and necessary decision.

The purpose of this chapter is to highlight the transformation (and inherent challenges) of Lincoln into a successful "STEM for all" school with particular focus on secondary grades of six to eight. This chapter will describe the process of designing a STEM program that meets the needs of all students, including students with disabilities and other diverse learning needs, as well as the challenges of establishing a STEM school within a large, urban district. The chapter will detail how a mandated NCLB school of redesign was transformed into an innovative, twenty-first-century learning environment. Details relative to teacher professional development (PD), curriculum design, and actual implementation will be shared.

Preliminary Work toward Opening Lincoln STEM School

At the time of its closure, Lincoln was like many other urban schools across the nation. While the student population was waning (around 150 students attending at closure when the school district projected a population of around 350), most of the students who attended Lincoln were from the surrounding neighborhood. Adjacent to the campus of Lincoln was a thriving public charter school. Many potential students went to this charter school because it was located in the neighborhood and had some relative academic success. Other students from the neighborhood selected to go elsewhere in the district to schools that were not on academic notice. The school was not only failing the students, it was also an economic burden to the local urban district. The school building was costing more to operate than what was fiscally responsible. Of the remaining students attending

Lincoln, nearly 99 percent were African American, and nearly all students came from families considered to be lower socioeconomic status (SES).

Also like many schools across the nation, the school's physical structure was in deplorable condition. Many walls and doorways in the school contained old, washed, or sloppy painted-over graffiti. Numerous windows were broken. Some were covered, by teachers, with plastic shopping or trash bags still allowing some weather to enter the building throughout the year. Heating units lay in pieces on the ground and covers to electrical sockets were broken and several were missing completely. The outside play area was nonexistent due to a large tree limb that had fallen sometime within the last year of operation and was still lying across the small space. The floors were in need of cleaning and waxing, and many walls were stained and needed paint.

Probably the most tragic room was the school's library, the cornerstone of any school. It had been closed due to the school's diminishing student population and district-wide budget cuts. The walls of the old library space were lined with old wooden bookshelves, some in splintered pieces, all lying empty. There were no traditional library books found on any shelves of the library. The remaining books, mostly old textbooks, were amassed in disheveled piles on the ground or on tables. Some books were stacked in shabby boxes on tables. There was a bucket and two old trashcans on tables that had been used to catch rain before the roof was repaired a year or so earlier. The combination heat and air-conditioning system was able to spew minimum heat. However the cooling unit, one of a few air-conditioning systems in the building, was beyond repair.

Behind the scenes, the STEM partnership saw opportunity in Lincoln's closure. It was an opportunity to transform a failing school into a twenty-first-century learning environment. Through the support of a state-sponsored competitive grant, the Cincinnati STEM partnership focused on gaining funding to reopen Lincoln as "Lincoln STEM School." It was envisioned to be the first prekindergarten through eighth-grade public school founded on a Universal Design for Learning (UDL), STEM initiative in the nation. This UDL-based STEM approach focused much of its interdisciplinary learning through problem-based learning and project-based learning (PBL). The vision was clear; to reopen a school that provided a learning environment for *all* students, a place to gain the knowledge and skills necessary for the modern world. The vision was to develop a *"STEM for all"* (Basham, Israel, and Maynard, 2010) neighborhood public school where students, families, and educators would work as a community with the goal of meaningfully engaging students through the completion of high school. As a school, it would require students from an early age to problem-solve, critically think, explore modern STEM careers, encourage the idea, and provide paths toward college. At a minimum, the school would provide the students with the twenty-first-century skills necessary for a self-determined life.

In the spring of 2008, funding for Lincoln's redesign was secured through a STEM Programs of Excellence grant, a competitive grant offered by the State of Ohio. It was in late April that a team from the UC was brought together to work with the school district and the lead school's redesign. The timeline was set for the new school to open in August of that year. So, the school would officially shut down in early June and reopen as a "new" UDL-based STEM school two months later. The task to rebuild the new Lincoln included hiring new teachers, providing them a month-long PD that helped them prepare this innovative curriculum, and procuring the necessary instructional materials and technology in order to implement the curriculum, in addition to designing and renovating the facilities.

The Birth of a New School

On a morning in late April 2008, a team from UC, the local school district, and other stakeholders came together at the central office to layout the logistics of reopening Lincoln as a "STEM for all" school. As an implementation team, we understood the task that we had been handed (and accepted) as very exhilarating. We also understood that while the task to transform the school was exciting, it was highly contested and politically heated within the community. In fact, throughout that late spring and summer, we were reminded on a number of occasions that success was essential, and that our jobs depended on it. The task we had accepted was inspiring, but appeared nearly impossible given the projected timeline. As defined, the task was to hire new teaching personnel, provide PD, and to redesign and renovate the physical structure of the building. A library of instructional materials and technology that would provide the necessary resources to support a STEM curriculum would also need to be purchased. Meeting our deadline to have the school ready by mid-August would be very challenging. By our April meeting, a pre-implementation team had already started to define the critical components of the school. Now, our team had to move from paper to reality. It was clear that two foundational underpinnings informed the redesign of Lincoln School into the Lincoln STEM School. First of all, science, technology, engineering, and mathematics would not be distinct subjects taught in isolation. Instead, these STEM areas would be considered interdisciplinary and be incorporated into other instructional areas including language arts, music, art education, and social studies. In this manner, students would learn that STEM does not exist only within the boundaries of traditional science classrooms, but it extends to problem solving in many areas of global importance. Second, Lincoln STEM School focused on *all* students' learning and academic development, regardless of their performance level. By beginning with the expectation that *all* students could benefit from and succeed in STEM learning; teachers and other professionals working at Lincoln were required to consider instructional planning that addressed instructional access, engagement, and accountability through the principles of UDL (Rose and Meyer, 2002; CAST, 2008) (to be discussed in greater detail later).

Interdisciplinary STEM Instruction

A fundamental element of Lincoln STEM is that STEM could not occur within the traditional silos of the specific content-area classes (e.g., science class). Instead, teachers within Lincoln had to consider that STEM was interdisciplinary, so students could begin to understand that science, technology, engineering, and mathematics go hand in hand and extend to other subject areas (Basham, Israel, and Maynard, 2010). For example, in order to design an effective bridge, students must not only understand basic mechanical engineering. They also need to understand the mathematical concepts behind the design, the behavior of different building materials used, and use general problem-solving skills. Students need to use technology to research the optimum design given particular constraints, and understand the nature of communicating this information to the community.

To truly begin to design a STEM PK-8 school, a great deal of collaborative planning, brainstorming, and problem solving needed to occur. All stakeholders involved in the redesign needed to partake in the discussion. This included the university faculty, the new school principal, school district administration, teachers, and union representatives. As a first step, configuration of the grade bands needed to be discussed. For example, what should a STEM school "look like" in primary grades? What should it look like in upper elementary school? What should it look like in grades six to eight? And finally, how do these grade bands relate?

Discussion of grade bands led to a major focus on curriculum. For testing purposes, the school district was strictly enforcing the notion of their own curriculum maps and pacing guides. They were reluctant to permit the implementation of "STEM" without adhering to these documents. Although some advocated for a "pure" STEM curriculum, the school district was unyielding with this request. Regardless, everyone agreed the content had to be taught through flexible curricular instruction and intermeshed with the district curriculum map, and modern instructional tools needed to be purposefully integrated into instruction.

"STEM for All" through Universal Design for Learning (UDL)

One tenet of Lincoln STEM Elementary is to provide engaging and challenging STEM instruction to *all* its students, not just to high-achieving students who traditionally are exposed to and thrive on STEM instruction. This school was different from other STEM-themed schools across the state and the nation, particularly because teachers had the primary focus on preparing all students with the necessary content and skills to fully participate in and benefit from the challenging and enriching problem- and project-based activities, design challenges, and other STEM-related instruction.

As teachers prepared for the students' arrival on the first day of school, one critical task was to design STEM instruction and curricular materials that would meet the needs of academically diverse learners. They did so through integrating the principles of UDL (Rose and Meyer, 2002) into their planning and instruction. UDL is a scientifically valid instructional design framework that incorporates multiple means of representation, expression, and engagement to proactively meet students' diverse learning needs (Higher Education Opportunity Act, 2008). To meet these content delivery, engagement, and demonstration principles, UDL rests on the meaningful use of both instructional strategy and technology that allows students to become more effective and efficient learners. In working with the teachers at Lincoln, UDL was taught and implemented through the use of backward planning (Wiggins and McTighe, 2005).

Consequently, content teachers, special educators, reading specialists, and other education personnel at Lincoln learned about UDL-focused lesson planning, how to identify and use high-impact instructional and assistive technologies, and how to construct lessons that allowed students to make informed choices in meeting their unique instructional needs. Additionally, once the school opened, teachers were provided with embedded PD to facilitate implementation as it related to daily lesson planning and problem solving. In this manner, teachers learned how to create engaging and accessible STEM learning that addressed the academic diversity within Lincoln STEM Elementary. As teachers became more proficient in integrating the principles of UDL into their instruction, their internal capacity increased, and they began to rely less on outside expertise from UC.

Day-to-Day Implementation Considerations

Beyond the foundational elements, the team had to deal with the realities of schooling in the United States. Since the implementation of NCLB, schools in the United States are not measured on whether they inspire a student to think, problem solve, or whether they provide for the diverse learning needs of students. Instead, the focus is on schools making adequate yearly progress (AYP) as measured by standardized tests. The current measures of school success provide an inadequate framework for measuring success in a STEM school (see Basham, Israel, and Maynard, 2010). Day-to-day operations of the school had to be much more pragmatic than having an interdisciplinary STEM focus that was based on UDL. On a daily basis the building principal and teachers ensured that Lincoln would be successful under the mandates of NCLB through AYP, regardless of the STEM education design. In order to meet all of the district requirements and the newly designed curriculum, the teachers needed an operational framework for carrying out UDL and STEM-based instruction.

Adequate Yearly Progress

While not the primary goal of the UC team or any other implementation team members, AYP was and still is a critically important success benchmark to many district personnel. As addressed in the NCLB Act, AYP serves as the measure of school and school district accountability for student learning (NCLB, 2001). While highly important to the school district, AYP was never the goal of the larger STEM partnership. The goal of making and maintaining AYP was and still is viewed as politically motivated and neglects the larger, more important goal of preparing students for life beyond the K-12 school system. The implementation team, along with the principal and teachers, had to continually find a balance between AYP and the desire to provide a richer set of twenty-first-century skills.

Problem-Based Learning and Project-Based Learning

In the first summer prior to the opening of the school, problem-based learning and PBL was a large focal point for the June month-long PD. Problem-based learning is situated around experiential learning organized through the investigation, explanation, and resolution of meaningful problems (Barrows, 2000; Hmelo-Silver, 2004). Many of the current instructional models used in medical and law schools are borrowed from the problem-based paradigm. With a slightly different focus, PBL was another framework used in the implementation of STEM instruction at Lincoln. PBL, used interchangeably with problem-based learning at Lincoln, involves learning experiences organized around a driving question (Blumenfeld et al., 1991) with the end result of a product. Thus with focus on engineering, PBL projects generally involve the development of a product. This product may be sophisticated as a mobile application or as simple as a paper-based model.

Problem-based learning and PBL were used at Lincoln because they were considered to be the best teaching practices that provide differentiation and readily integrate the multiple disciplines of STEM (Basham, Israel, and Maynard, 2010). As part of PBL, K-12 teachers and students are encouraged to work with the university, industry, and business partners as a means to strengthen resources necessary for the implementation of these innovative pedagogical techniques. Holistically, it is believed that working with STEM partners provides students with a more complete understanding of STEM careers.

Making It Accessible and Usable

Lincoln STEM Elementary focuses on implementing STEM instruction in a manner that is accessible, engaging, and challenging to students with diverse academic needs. This is done through several mechanisms: UDL integration, use of instructional technology, and data-driven decision making. All three of these elements work together and address the wide array of students' instructional needs.

UDL offers an instructional design framework for providing accessible, engaging, and meaningful instruction. It is designed to support students' academic diversity from instructional design through assessment. Teachers at Lincoln lesson plan along the three principles of UDL: multiple means of representation, multiple means of engagement, and multiple means of expression. By proactively considering how knowledge can be represented and demonstrated, the teachers can address the academic diversity within their classrooms. For example, students may access content through traditional text, text-to-speech software, or multimedia avenues. Similarly, they can express their understanding through multiple means such as traditional reports, poster presentations, and multimedia presentations. In this manner, students become empowered to make instructional decisions based on their needs, skills, and motivating factors.

The focus on UDL directly influences the types of instructional technologies and strategies used to present content and engage students in learning. Instructional technologies can make text accessible through text-to-speech software such as Read Please and Google Translate for language translation. They can also provide meaningful and engaging background knowledge through multimedia presentations using iTunes U and assist with writing tasks via Google Scribe. For example, PBL projects using Digital Backpacks (Basham, Meyer, and Perry, 2010) allow students to explore and share their knowledge. With Digital Backpacks, students have access to iPads, cameras, sensors and probes related to their specific areas of inquiry, and other technologies that enhance their learning and allow them to creatively demonstrate their understanding. It is through this powerful framework, instructional strategy, and technology that teachers are able to support struggling learners not only in accessing and understanding critical STEM content, but also in manipulating and extending that knowledge.

As mentioned, one example of intermeshing UDL, STEM, and PBL is through the use of Digital Backpacks. This simple design allows students flexibility in how they come to understand subject matter, engage in content, and express understanding. Through a simple instructional design framework of the Digital Backpack (see Basham, Meyer, and Perry, 2010), teachers are able to quickly design Digital Backpacks to meet the needs of diverse learners. For instance, one of the Digital Backpack designs that has been used at Lincoln STEM is a backpack outfitted for doing fieldwork with fossils. This backpack contains a laptop with a camera, a video camera, a ruler, an iPod touch, and a microscope. Students use videos and other documents to learn about fossils and then use the other mobile components to carry out the instructional task of gathering data and identifying fossils. Based on student needs, they can access the information in a variety of ways: reading, watching a video, or problem solving with friends. Then, the teacher can use a variety of instructional strategies to scaffold student understanding and support students in collaborative groups.

The focus on UDL and instructional technology only occurs because the teachers at Lincoln STEM Elementary are committed to data-based instructional decision making. All students are assessed to ensure that they master content through short-cycle assessments. Information gathered through this meaningful and purposeful assessment process informs teachers about students' needs for additional support and/or enrichment. The assessments also inform the administration about school-wide academic trends that may lead to changes such as shuffling resources and specific instructional strategy needs.

Moving from Paper to Reality

As previously stated, in the summer of 2008, the implementation team had to move from paper to reality. That summer, the team's task was to assist the administration with the hiring of new teachers, preparing the teachers for instructing in the paradigm of STEM/PBL and UDL, updating the physical school and grounds, as well as providing the instructional materials and resources necessary for work in this new environment. Each of these aspects required both time and energy. In this section, we share a snapshot of some of this work and the challenges and turbulence encountered with implementation.

Preparing Teachers for Lincoln STEM

In June 2008, the newly hired teachers participated in an intensive 22-day PD on the critical elements of UDL, STEM, and PBL. Led by the UC team members, the PD integrated these critical elements into instruction by introducing the teachers to the new school curriculum design, and allowing them to envision what a new STEM/PBL school might entail. After these introductory sessions, the teachers began to coalesce as a unit, building and expanding upon each other's strengths. This was an important starting point to the PD and the "ownership" of Lincoln, as strong unity sustained the teachers through challenges encountered later in the school year. Situating the teachers in a collaborative leadership role to support their own understanding of UDL, STEM, and PBL was another key element in the cohesive nature of the PD. For example, one teacher, Rob, recognized that co-teaching, known as a process where two teachers work together to instruct (Murawski and Swanson, 2001), would be useful in a STEM environment. He asked if he could guide his new colleagues through the understanding of co-teaching models. The next day, Rob led a short presentation, discussion, and activity about how co-teaching could be used in the school.

Each day, the PD began with a recap of the previous day's and evening's work. The UC team would facilitate these sessions and present the deliverables that needed to be accomplished by the end of that PD session. After these daily instructional periods, small breakout groups of teachers would work on planned activities and tasks to achieve these goals. For example, if the goal of one day was to focus on the development of PBL activities across grade levels, the teachers in that grade level would work to align their school district–required curriculum maps with newly designed PBLs and/or other activities. If group members were confused with a task, they would stop and ask the larger group to help them with their problem. This process built a strong and cohesive team that was accepting of each other as individuals, as well as professionals. Another emerging result of this process involved the teachers' value of collaborative problem solving as a way to overcome challenges. By providing the teachers with the necessary time and tools, they felt a sense of ownership toward their school. Realizing the importance of teacher empowerment, the UC team continued to facilitate this process throughout this month-long PD.

Another important component of the daily PD routine involved the use of a "parking lot" list. The "parking lot" list consisted of a list of concerns that the teachers had throughout this process of redesigning the school. Using this strategy, anyone could add items to the list at any time throughout the day. Items that appeared on the parking-lot list could be anything, either small or large. Generally, these items could not be immediately dealt with (at that given time) for a variety of reasons, but were important to address when at a later time. Most of the items in the parking lot were not in the teachers' direct control. For example, some of the items included resources for their classrooms, funding, collaborative time to meet during the school year, and so on. Routinely during the daily debriefing, parking-lot issues were revisited, added to, dealt with, and/or crossed off the list. At the end of each PD session, teachers were asked to reflect on issues and concerns that they encountered during the day.

Generally, the parking lot provided the UC implementation team discussion topics and an action plan for work with district departments and contractors. For example, a parking-lot topic that arose centered on making sure that the teachers and students would have wireless Internet access when the school district did not support wireless access. Availability of needed instructional tools should not be teachers' concerns, but this was a major focus for the implementation team. The implementation team maneuvered district barriers to provide wireless access to the Internet. Examples of other parking-lot items included a requirement on the school district's mandatory teacher observation form that all students had to be seated and working rather than standing and being physically engaged in activity such as PBL (where students would be standing and moving around). In both these examples, the UC implementation team negotiated with the school district on behalf of the teachers.

Another large focus for the initial PD was the introduction of PBL through the lens of UDL and STEM. This introduction created cognitive dissonance with the newly hired Lincoln teachers. Each of the teachers was hired because they expressed an interest in innovative pedagogy and inquiry, both elements required for the integration of UDL, PBL, and STEM into their existing curriculum. To begin this process of shifting paradigms from the teaching and learning that they were used to in their old schools to the integration of UDL and STEM (via PBL) into the curriculum required the teachers to realize the resources and partners they had available to them. The UC implementation team took the opportunity to introduce the teachers to the existing resources and partners as a means to help them to understand and appreciation how STEM and PBL could be introduced in their classrooms. For instance, major business partners included the Cincinnati Zoo and Botanical Gardens, GE Aviation, and the National Underground Railroad Freedom Center. Our belief was that the long-term process of establishing sustainability and scalability of the STEM initiative was based on the fundamental strength of our partnerships and the regional network. This process of collaboration among the partners continued throughout the project.

Updating the Physical Infrastructure

As previously discussed, the physical spaces within the building were in dire need of renovation. This process began by creating a punch list of required renovations to the building. This was accomplished by completing multiple walk-throughs of the school with various district officials, groups (e.g., director of facilities, the building principal, etc.), and contractors, noting where changes were needed prior to the start of school. At the end of this process, a 20-page list was developed of items required to be addressed before the start of school, and another 8-page list of desired, although not imperative items, was requested.

To support the work at the school, a school district facilities intern (from the university) was put in charge of the facilities update project. Starting in July, facilities and instructional technology and various other departmental personnel (e.g., electrical, etc.) started the school renovation. Given the timeline of less than six weeks before the start of school, the process of "swarming" the building was used. During "swarming," the project manager assigned a number of individuals and groups to work at the building at the same time. Along with district facilities and IT, various community groups also donated time and supplies to help with items such as painting and grounds landscaping. For two weekends, volunteer groups worked tirelessly on sprucing up the facility. In the end all 28 pages on the punch list were completed.

Providing the Needed Instructional Tools

Beyond preparing the teachers and updating the school's infrastructure, efforts were made to anticipate both student and teacher needs in this new UDL-based STEM environment. Importantly, we focused on "instructional tools" rather than "technology," because technology conjures up thoughts of computers, keyboards, and if we're lucky the Internet. In the summer of 2008, our broad definition of instructional tools caused a few disagreements and meaningful discussion around the question, what does the twenty-first-century school have in it to enhance learning? What tools are needed to support instruction in a UDL-based STEM school? And importantly, how do we define those objects in the school? We concluded that "instructional tools" are any tool used to support instruction. This definition included what has been viewed traditionally as technology along with the lab equipment, science kits, and supplies for supporting specific learning. In this sense, no individual (science teacher) or department (e.g., math, special education) limited the use of an instructional tool due to some arbitrary rule. Any of the tools, for example, student laptop carts (all teachers had laptops), USB microscopes, digital probes, or iPod touches could be used by anyone throughout the building—one simply needed an instructional purpose.

The general focus for this early stage of outfitting the learning environment involved developing an acceptable foundation as well as library instructional tools for a twenty-first-century school. Through this initial purchase, we provided both distributed as well as targeted needs. The key was to focus on tools that provided the greatest usability and flexibility across the environment. The joint implementation team made tool acquisition focused on the key elements of UDL, STEM, and PBL. The newly hired teachers played various roles in the purchasing process; at times they served as leaders, suggesting solutions, at other times they served as users, providing feedback to the team. Initial selection of tools addressed primary needs, with a great focus on accessibility associated with digital learning objects (e.g., digital text, movies, simulations, etc.). For instance, each room was outfitted with an interactive whiteboard system. Then based on the desire for flexibility, we purchased multiple laptop carts (including Apple MacBooks and installation options of Windows, again focused on flexibility), an iPod touch cart, and various small tools (e.g., USB microscopes, data-collection probes, video cameras, still cameras). Then, actual backpacks or bags were purchased to support students taking these tools beyond school grounds. Thankfully, the district had already made an investment in science kits and other essential curriculum-based materials (e.g., math manipulatives). This initial purchase encouraged teachers to use digital materials within their lessons.

Importantly, the team did not design the school to have a computer lab. In a purposeful decision, the computer lab was viewed as providing a separate place to do technology and this was against the greater vision of "STEM for *all*" students across the curriculum. In fact, an isolating computer lab parallels the notion of science only taking place in the science lab or math only taking place in the math classroom. Again, the idea was to break down the walls of the silos. Following the principles of UDL and a modern STEM school, technology is used seamlessly as an instructional tool throughout the day in every environment. The school also has targeted learning environments.

One example of a specifically designed location was the old library, a place designed for interdisciplinary and collaborative learning. Affectionately termed the "Digital Commons," the design of the room encourages collaborative work among both students and teachers. In a simple design, the Digital Commons contains five interactive whiteboards for students to collaboratively work on PBLs and for teachers to work on learning teams or attend a PD workshop. For students, a team of students can sit in a horseshoe desk around one of the interactive whiteboards researching a topic, developing a solution to a PBL, or developing an artifact or presentation to share with the class. For teachers, the Digital Commons provides a location to design PBLs or to look at student data that assists them to make decisions about instructional design.

Now in its third year of operation, the focus continues to be on flexibility. Specifically, teachers have open conversations about their most-prized usable and flexible tools. For instance, many have noted that the iPod touch is one of the most flexible instructional tools in the school (at the time of this writing they had just received a few iPads). The iPod touch can be quickly (and inexpensively) uploaded with needed movies, podcasts, and apps, and then deployed to students, for individual or small-group work on any number of learning activities. Finally, while the Digital Commons is still operational, Lincoln now has a computer lab of donated new computers. This lab, designed to be used on weekends and evenings, provides for community outreach events and a place for parents to come in and learn how to use specific pieces of technology or how to complete tasks (e.g., build a resumé). During the day, this computer lab supports students working on projects. It also provides students with an extra place to work after school as well as provides the teachers a way to continue covering the curriculum.

Providing the Instructional Resources

The resources to make Lincoln STEM operational moved beyond the need for "cool tools" and required an effort to support instruction in a variety of ways. Limited instructional time creates a barrier for teachers, and at Lincoln, this is no exception. One of the early parking-lot items identified by teachers was the need for collaborative time for instructional planning (e.g., more time to make critical decisions around student data, time to support student needs). While the implementation team could not actually provide more time per se, we anticipated this need and provided a graduate instructional designer to support UDL, STEM, and PBL in the school. The graduate assistant (GA) worked four days per week supporting the teachers' technology needs. In the first few months, the GA provided the school with day-to-day instructional troubleshooting, some technology support, and some instructional design, but in January of that first year, her duties shifted toward mainly supporting instructional design. As the instructional designer, she would identify potential instructional tools (e.g., Google maps) that could be used during an instructional session and assist teachers when they tried to implement this technology in lessons and units (module designs). In addition to these responsibilities, she assisted teachers with actual instructional implementation in the classroom for students.

In addition to providing the school with the GA, different members of the implementation team were also present at the school for an average of a day a week. The team worked with the GA, teachers, and other staff members in their role to support student learning. The roles of the implementation team also facilitated individualized and small-group embedded PD to support the needs of teachers. This embedded PD included topics such as troubleshooting the interactive whiteboard and other technology, using iMovies, supporting students with low reading comprehension, and designing a PBL to meet the learning standard on habitats. Through this first year, the teachers operationalized expectations for what instruction looked like within their school. They became confident in their ability to use the technology and to operate under the UDL and STEM framework.

Lessons Learned from Implementation

Providing challenging, high-interest STEM instruction to students with diverse academic needs is possible through proactive instructional planning, appropriate curricula, and instructional technologies. It, however, does not occur in isolation. Over the last few years, many challenges have emerged and many lessons have been learned. Table 1.1 provides a snapshot of initial challenges and strategies used to overcome the challenges. A more detailed description of each of the challenges and how we maneuvered through them is provided. We hope that the challenges we encountered

Table 1.1 A Snapshot of Initial Challenges and Strategies

Category	Challenges	Strategies
Technology	School district vision for district-wide instructional technology limited -to MS-Office.	Worked with partners, the teachers, and other innovators to make an initial investment in the technology. We then provided a GA to help support instructional planning and instructional aide.
Curriculum	Prescribed by school district; used "canned" curriculum as the norm, all benchmark and pacing guides for the school district implemented in Lincoln; PBL/STEM instructional materials were limited; aligning PBL curriculum with school district requirements has been difficult; no time for teachers to collaborate.	Grant funding changed the direction of curriculum toward using PBL framework; use of Engineering is Elementary (EIE) as framework for STEM; ongoing PD trains teachers in lesson-plan design and implementation; support for teachers in the classroom is essential.
Community	Poor reputation in school district; considered a "failing school" per NCLB; parents and community were distant from school.	The neighborhood and community began to rally around Lincoln when they witnessed the changes that occurred with students and surrounding area. Teachers and others attended numerous community events.
Facilities	Location of school was undetermined until June; building chosen was in horrific-condition graffiti, dirt, inoperable shades and furniture, tripping hazards, clutter, broken windows replaced with blue plastic.	University faculty took initiative to maneuver the system to get Lincoln in working condition by the start of school, time lines developed; pressure on school district to push agenda to open school, technology in place for the start of school. Used what we called "swarm" where various groups were working on the building at the same time.

Continued

Table 1.1 Continued

Category	Challenges	Strategies
Teachers	All teachers in the school needed to be hired prior to the start of school, but not all were hired by that time; several teachers have left the school prior to Winter break because they were overwhelmed with challenges of the school; work was extremely difficult for teachers.	Core teachers ($n = 7$) is a strong and cohesive group; teachers work extremely hard to make the school a success.
Professional Development	New teachers were hired without working knowledge of PBL/ STEM framework; not all teachers were hired for initial PD; teachers' union protective of teachers' time for PD, concurrent PD ongoing during June was mandatory for teachers.	PD took place during the month of June; established PBL/STEM framework for lesson development; initial PBL lessons were designed; embedded PD during the first half of the school year; technology training ongoing.
Student Behavior	Students had the reputation of "running the school" in the past; behavioral issues; lack of respect for the school.	Students want to learn when they are engaged. Lessons and projects engaged the students.
School District	Various complexities of a large school district are apparent here, many issues were difficult to maneuver; school district pressured school to increase enrollment; strong teachers' union made it sometimes difficult for both school district and university personnel.	Transparency with school and district was the biggest asset to help develop Lincoln. Reporting directly to the president of the University, as well as the superintendent sometimes provided a big stick to get things done.

and the strategies we used to overcome them can provide a framework for others presented with such situations.

Technology

The challenge we faced was developing a twenty-first-century learning environment, with a district that had a limited understanding of how technology related to curriculum and instruction. Likely due to this lack of understanding, the district also had a very limited budget for instructional technology. The school district's base technology included five semimodern computers per room with MS-Office installed. Through the STEM grant, we purchased mobile computer carts (we did not have the money to purchase one laptop per student), interactive whiteboards, wireless Internet connections throughout the school, digital document projectors, handheld electronic devices such as the iPod Touches, digital still and video cameras, USB microscopes, and the software and teacher training to support this endeavor. To maneuver through this challenge, we sought to work with business partners and other innovators to provide support for this initial investment in the technology. We provided the teachers with the skills and knowledge to meaningfully use the technology during UDL-based STEM instruction.

Curriculum

Like many districts, the curriculum was prescribed and set to a curriculum map with pacing guides that provide teachers with the specifics of what should be taught on a day-to-day basis. This "teacher-proof" canned curriculum along with quarterly benchmark testing helped the district ensure all students throughout the district would be prepared for end-of-the-year AYP testing. The implementation of a STEM theme in a failing school was met with some resistance from some individuals in the school district. While nearly all encouraged the twenty-first-century skills, the pedagogical approach, and even the heightened STEM content, many feared the extra time and the needed curriculum flexibility it would take to actually implement. Essentially, it left too much in the hands of the teachers and possibly not enough time focused on the "TEST." Thus, the teachers were pushed to maintain alignment with the pacing guides, which provided little flexibility to implement this new design.

Focused on pacing guide alignment, the teachers continually noted how frustrating and difficult it was trying to align STEM (PBL) activities with school district curriculum requirements. Moreover, the school district provided little time for instructional planning and proactively designing a flexible STEM-based curriculum. The teachers constantly needed to attend after school "mandatory" school and district administrative meetings. To negotiate these challenges, the grant provided funds to buy new curriculum materials, thus changing the direction of curriculum using PBL framework. The purchase of the Engineering is Elementary (EIE) curriculum, while not perfect, provided them a framework for thinking about STEM. This curriculum provided teachers with guidance as to what STEM activities involved. The implementation team then provided embedded PD that educated the teachers in UDL- and STEM-based lesson planning. The important aspect was the support provided for teachers in the classroom. This support came in a variety of ways and gave the teachers encouragement that they were not alone in this endeavor, it was a school-wide initiative, and everyone embarked on the journey together.

Community

Because of past experiences, many in the community surrounding the school mistrusted both district and school personnel. Early on, we decided that if the school transformation were to be successful, we needed to gain community support. We initiated various efforts throughout the initial summer and the following school years to gain and maintain community trust. The initiative was started by organizing efforts with local churches and organizations to help beautify the building and the school grounds. Early in the first summer, implementation team members, school staff, and other community volunteers worked hand in hand to remove trash, clean up graffiti, and renew the landscaping. Next, implementation team members as well as the newly hired staff attended a number of church barbeques and various events throughout the community to engage parents and community members and discuss plans for redesigning the school. Along with all of these initiatives, the district hired a community-centered principal. This individual rallied the neighborhood in several different ways. First, she and the teachers met with the parents multiple times before the beginning of school year, as well as several times during the school year, many times on Saturday mornings. During these meetings, she would personally cater the events with food and beverages. She created an "open-door" policy for parents to visit the school to volunteer in the classrooms and throughout the school. Second, she was a neighborhood advocate and contacted an award-winning hospital (located in the community) to partner with the school. The hospital provided free nutrition-education services as well as other free health services. Third, when needed, she helped parents obtain required necessities (e.g., winter coats and food) that were not always available to the students at home. Finally, she held community events at the school, from providing weekend computer-education classes to holiday events for parents to see their students perform. In time, parents became more active and trusting of the school. Today this change is still evident by the school's active and growing parent-teachers organization.

Facilities

As previously discussed, the school building itself had fallen into a state of disrepair. A major focus of the implementation team was to provide the students with what felt like a new school without having the funds to actually construct a new school. Fortunately, the school district, the university, and the various partners also wanted an updated school for the students to attend. The leadership within the school district encouraged the implementation team members to develop a "punch list" of items that needed to be fixed and updated. Targeted items focused on safety and infrastructure, and then we turned to supporting a twenty-first-century learning environment. All updates to the actual learning environment were made with flexibility, accessibility, and budget in mind. When the district was unable to commit to a fix or update (e.g., painting the rooms), the implementation team would turn to the project partners and community. Then as a team, a plan was developed for the completion of the list. Through much coordination among the district, the university, and the project and community partners the implementation team along with district leadership organized work crews to swarm the building, providing large teams of people focused on specific interrelated tasks. Vastly different from the way the district normally operated, this approach allowed the revitalization of the building to take place in two to three weeks.

Teachers and Professional Development

Because of the school's "redesign" designation, all previously employed teachers at the school were dismissed, which required that a new staff had to be hired. Members from the UC implementation team attended "hiring sessions" where prospective teachers participated in informational sessions about the new school design. Although the implementation team was not part of the decision-making group for hiring teachers, the team provided some insight as to potential hires. We believed that the ideal teacher candidate for the school needed to exhibit reform-minded thinking about teaching and learning. Following the union protocol, teachers were hired based on what "hiring-round" they applied in, and by June 1, only six new teachers hired to the school, leaving nine positions still open. This was of concern because by the third round (starting June 1), many of the best teachers had been picked over in rounds one and two. With the school scheduled to open on August 18 and with the start of the PD on June 3, it became a bleak prospect for the school. Regardless of this challenge, the PD began as scheduled on June 3. Not surprisingly, all of the teachers hired lacked a working knowledge of UDL, STEM, or the PBL framework.

The core group of six became a cohesive unit and worked extremely well together. As new members entered the school (through hiring), the core group extended a welcome to the new teachers. Unfortunately, it became apparent that although the additional round of new teachers was welcomed by the group, they were not always amenable to the new way of approaching teaching. Non-reform-minded teachers distanced themselves from those who were more reform minded. This caused friction among the staff, and eventually nearly all of the teachers not open to a new way of teaching left the school prior to the first winter break (causing position openings). Generally, these teachers expressed that they were overwhelmed with challenges of the school (e.g., long working hours, continual press, people walking in and out of rooms, difficult student-behavior issues, weekend responsibilities, and mandatory meetings everyday after school). Teaching vacancies were filled as quickly as possible, and newly hired teachers promptly received support to integrate into the value structure of the school.

Student Behavior

As expected in the failing school, students exhibited maladaptive behaviors such as being off-task, and problem behavior was overwhelming. With a lack of meaningful engagement and increased student behavior problems, the students were not learning. In fact, the school had a reputation of students "running the school" and lacking respect for the staff. Beyond being a community-centered principal, the school leader displayed a strong but respectful discipline style. The school enacted a data-based Positive Behavior Intervention Support model that the district called Positive School Culture. The principal and staff then dedicated the first several months of the school year to changing the culture of the school. This included procedures and rules such as: (a) no students could roam the halls during classes; (b) no talking in the hallways; (c) all students were expected to be quiet when teachers requested attention; (d) students needed to line up when walking in the halls; and (e) no tolerance for fights, disrespect of anyone. Whenever incidents emerged that required the principal's attention, she would quietly, yet strongly and respectfully intervene.

The principal was present in the building at all times, walking the hallways, talking to the students, making them feel wanted and loved. Her strong, yet gentle demeanor comforted many students, and she projected a genuine caring of all students in the school. She and everyone in the school modeled appropriate behavior, and the students learned school expectations. After three months of modeling appropriate behavior, the school culture began to change. Lessons and projects were interesting and engaged the students. The students then demonstrated a true desire to learn. The culture created resulted in students viewing school as an exciting and privileged place to learn. As a result, test scores increased by 30 percent from the prior year district benchmarks during the first half of the year. Absenteeism decreased, and enrollment began to increase. As parents heard of the success of Lincoln, they wanted their children to attend this promising school.

School District

Like many urban school districts, the organization itself offered challenges to redesigning the school in such a short period of time. For instance, numerous departments and offices within the district lacked an understanding of the project and the team. When we requested a meeting with groups (e.g., facilities), it was necessary to introduce and explain the project. The lack of communication among departments produced power struggles and arguments among district personnel. These arguments generally resulted from someone feeling left out of the decision-making process. To lessen the conflict, we encouraged complete transparency in everything done throughout the project.

Moreover, the district included individuals who had both open and hidden agendas, as well as politics not always centered on providing the best learning environment for children. These landmines had to be negotiated or set off based on the situation and need. Luckily, the project had both high and deep community partner support. For instance, when a department was unwilling to do something (e.g., move old furniture from the building), the implementation team communicated to the unwilling party how we would have to move up the organizational, community, or political food chain. Thus, we most often found that necessary implementation steps would occur.

Opening Day and Beyond

After one busy summer, the school reopened on time, August 18, 2008. Now, three years after the redesign, the school is considered a success. While the implementation team disbanded (some have moved onto other jobs), most of us on the implementation team still serve as formal and informal advisors working with individual teachers and teams at Lincoln STEM. Team members from UC still meet with teachers, generally informally, to discuss instructional designs and solutions to problems that might be occurring at the school. We are also now involved in various implementation and research projects at the school. Over the years, we learned various lessons about developing this "STEM for all" environment. Here are a few of the lessons learned:

- Building a strong community around a school's transformation and sustained operation is critical. Having a university and school district partnership is a very beneficial team.
- There must be continued commitment to programmatic sustainability through:
  - Ongoing PD and support for new teachers entering the school. As new teachers enter Lincoln, they must learn how to create effective interdisciplinary instruction that embeds the principles of UDL, STEM, and PBL.
  - There must be a mechanism in the school that acculturates new teachers into the academic culture within Lincoln.
  - There needs to be a system in place that allows for embedded PD to allow teachers to work through specific UDL-based STEM instructional challenges. Even teachers who are proficient at providing meaningful interdisciplinary STEM instruction that addresses the needs of diverse learners occasionally meet instructional challenges for which they can benefit from ongoing support. Isolated, onetime PD may provide a good overview of effective instruction, but it cannot address the specific needs of teachers. Consequently, ongoing, embedded, and just-in-time PD addresses the unique challenges faced by teachers as they try to meet their students' needs.
- Concepts and frameworks such as UDL and STEM must fit within the global-reform initiatives of the school district so that they are not seen as another thing on top of other reform efforts. Because school districts constantly change reform efforts, in order for meaningful school change to occur as it relates to effective STEM instruction, the components of that instruction (i.e., interdisciplinary instruction, UDL) must purposefully be integrated into the current reform efforts of the school. This takes a great deal of collaboration between district curriculum leaders, school administrators, and teacher leaders.
- There must be an initial and ongoing financial support that allows for purchasing of modern instructional technologies and curricular materials as well as support for teachers' PD. Because many schools do not initially have sufficient curriculum materials and instructional technologies, it is important to clearly understand the financial needs of the school.

In Closing

As of this writing, the school is midway through its third year of operation, and while not the in the clear, it is heading down a path of success. We can ensure that, provided the appropriate resources and motivation, the groundwork can be laid for schools to transform in relatively short amount of time. In our situation, we began to see change within the first week when behavior problems decreased and has remained consistently better (98 percent of all problem behavior is dealt with in the classroom). We then saw student performance increase. Within the first three months of school, the teams saw a 30 percent gain in quarterly assessments from year prior (with the same population of kids).

Primary to Lincoln's early transformation are the various partners, the basic financial support, and teacher support. If any of the key variables changed, the outcomes of this complex equation likely would have changed. We have concluded that there is no magic pill or practice in education. Within our case, we were able to quickly develop the infrastructure (both human and physical) in a two-and-a-half-month period, however the real work was done day in and day out in the classrooms. To this day, the daily work requires a committed group of personnel (teachers, building administration, and staff) who feel rewarded as well as supported by district administration. Moreover, these personnel must continue to receive the necessary PD around STEM knowledge, skills, instructional tools, and resources for supporting this interdisciplinary twenty-first-century learning.

Chapter 2

The Importance of Up-Front Evaluation Planning Including Student Learning Outcomes

Shelly Sheats Harkness and Catherine Lane

Professional development is hard work. It cannot be taken lightly if it is to make an impact in schools, with teachers, and ultimately on student learning. Within this chapter we report on the lessons learned from our participation in a National Science Foundation (NSF)-funded professional development program, *Mathematics Teacher Leaders* (*MTL*, pseudonym). The professional development leadership group first recruited local site teams at universities in three different regions (Midwest, Northwest, and Southwest) for the *MTL* program. The three local sites were comparable in that they committed to work with culturally diverse school districts identified as "urban, urban... predominately African-American, and predominately Hispanic" (program documents). Local site team leaders selected high schools in their districts and met with teachers and administrators in these school districts to begin to develop specific plans for *MTL*. Subsequently, these school leaders were asked to create plans for professional development that would meet each individual school's needs and ultimately affect teachers' practice and student outcomes.

Our work with a large urban school system in the Midwest was deemed "unsuccessful," and we lost NSF funding after two years of implementation. In a previous article (Harkness, Plante, and Lane, 2008), we used the *Ohio Standards for Professional Development (OSPD)*, contained within the *Standards for Ohio Educators,* to rate our efforts in this same program. We discuss not only these ratings in this chapter but also delve deeper into other issues that impacted our professional development efforts. We focus on the importance of creating an evaluation plan at the beginning of the planning process using Guskey's (2000) "Critical Levels of Professional Development Evaluation" as a framework. As Guskey noted, this evaluation plan should include measures that assess the expressed goals and also *articulate* and *define* student learning outcomes. We begin by addressing two questions: Does professional development make an impact? Can professional development make an impact?

According to Stigler, data from the evaluation of professional development programs indicate that it "doesn't really help teachers or students learn more" (Willis, 2002, p. 6). Reasons abound that tend to support Stigler's statement. Schools are notoriously characterized as deprofessionalized "egg carton" structures (Lortie, 1977). Teachers work in their own classrooms, secluded from what happens outside their closed door "comfort zone[s]" (Schoenfeld, 2009). Additionally, "... low-trust, blame cultures abound in public services" (Knight, 2002, p. 240). Blame for students' low test scores or low achievement is typically passed from students to parents to teachers and to administrators. These structures and cultures must be changed, and collaboration must become a reality in order for professional development to make an impact.

Furthermore, vanDriel, Beijaard, and Verloop (2001) note that reform efforts in the past have been unsuccessful because the professional development providers failed to take into account teachers' existing knowledge, beliefs, and attitudes about teaching and learning. Typically, the role of teachers in reform movements has been one of "executing the innovative ideas of others (policy makers, curriculum designers, researchers, and the like)" (p. 140), and, accordingly, this type of professional development is doomed to fail. What happens is that teachers tend to "tinker" with new materials and techniques, and then merely incorporate these into their existing practice (Thompson and Zeuli, 1999) without changing their beliefs and attitudes.

Despite these gloomy depictions, we believe that professional development can have an impact. Clarke (1994) denoted at least three areas that professional development can impact: (1) "establishing" to promote organizational change, (2) "maintenance," and (3) "enhancement" to improve individual teachers' practices (p. 37).

In fact, we remember our days as classroom teachers and our own experiences as consumers of professional development. We concur with Darling-Hammond (1998) in her notion that students benefit from their teachers' opportunities to learn. Most of our experiences probably fell under the "enhancement" impact (Clarke, 1994). Some experiences were better than others. However, we cannot say definitively what made some better than others or what impact they had on our teaching or on our students' learning. Because of these reasons, we searched literature about the impact of professional development and then, more specifically, literature related to the evaluation of professional development. The reviews of these searches follow.

Professional Development

As early as 1991, the National Council of Teachers of Mathematics (NCTM) recommended that school administrators should assist teachers by engaging them in the design of professional development specific to their needs. More recently, in the Sixty-Sixth NCTM Yearbook, *Perspectives on the Teaching of Mathematics*, Taylor (2004) described collegial interactions—defined as collaboration among colleagues that is focused on common goals—that can enhance the "growth" of mathematics teachers as a critical need for professional development programs. Specific kinds of collegial interactions within professional development programs that lead to teacher change were characterized (Little, 1982):

1. Teachers engage in regular, constant, and increasingly concrete and precise talk about their practice.
2. Teachers are regularly observed and provided with useful evaluations of their teaching.
3. Teachers plan, design, research, evaluate, and prepare lessons and other materials.
4. Teachers teach each other the practice of teaching (p. 331).

Although we cannot argue that these interactions are unimportant, what seems to be missing from this list is any mention of students. Teachers should be talking about the impact of practice on their students' learning (see 1 above). The observations should focus on what students are doing while the teachers are teaching (see 2 above). In fact, Wilson and Berne (1999) included students in one of their three "knowledge" acquired via professional development categories; teachers should be provided with opportunities to talk about: (1) or do subject matter; (2) student thinking and learning; and (3) teaching. Additionally, we wholeheartedly agree with Wilson and Berne that subject matter should be another focus for professional development in mathematics at all levels (secondary, middle, and elementary).

A review of professional development programs that demonstrated evidence of improved student learning by Kennedy (1998) found that these programs focused on helping teachers learn how students learn the subject matter. The teachers also learned: the subject-matter content in deeper ways; how to recognize if and how students were learning; and specific pedagogical practices (Kennedy, 1998). Based on their review of literature, Thompson and Zeuli (1999) citing others' research (Brown, Collins, and Duguid, 1989; Collins, Brown, and Newman, 1989; Huberman, 1995; Ball and Cohen, 1999) recommended professional development programs must:

1. Create "cognitive dissonance" to unsettle the balance between teachers' existing beliefs and practices and their experience with subject matter, students' learning, and/or teaching.
2. Provide time and contexts for teachers to grapple with the dissonance through collaborative discussions.
3. Make sure that the "dissonance-creating" and the "dissonance-resolving" activities are connected to discussions about the teachers' own students and classrooms.
4. Help teachers develop a collection of practices that meshes with their new beliefs.
5. Make this process cyclical (p. 263).

Also focusing on what works in professional development programs, Birman et al. (2000) assessed the impact of Eisenhower Professional Development Program grants and found that professional development should feature the following: deepening of teachers' content knowledge and knowledge of how students learn that content; opportunities for active learning (to observe and to-be-observed teaching, to plan lessons, to review student work, and to present, lead, or write); coherence in experiences (build on earlier activities and involve teachers in discussions with each other); duration; and collective participation.

Furthermore, a strong principal and a strong superintendent are critical stakeholders in professional development programs. Professional development that works includes:

> Ongoing presence and involvement of liaisons from the university faculty who share a mission with the school and its staff... paths of college faculty crisscross so regularly with the daily routines and expectations of school teachers, students and administrators that interchange and mutual support are standard; college professors become trusted colleagues rather than idealistic, "clueless" interlopers. The professors become an integral part of

> the professional development school, bringing with them new ideas, techniques, research findings, and pre-service teachers. School life ultimately engages the professors, imbuing them with a sensitivity to teachers' professional contexts (McBee and Moss, 2002, p. 64).

Additionally, the person who leads a school department matters. "In the long run a good way to become good at leading to learn is by being in communities in which it happens" (Knight, 2002, p. 236). More succinctly, department chairs model professional learning when they participate in learning communities. Knight made five suggestions for department heads: (1) publicize the learning message; (2) be a role model; (3) evaluate for learning; (4) clarify what subject knowledge is for... use big ideas that define the subject as taught in a department; and (5) assume all children can be learners.

Bullough and Kauchak (1997) studied the school and university partnerships in three secondary schools. They highlighted their work in secondary schools because very little research has been conducted in secondary schools and because of the unique challenges faced by these partnerships. They wanted to know if the partnership work "influenced teachers' thinking and practice, changed schools, and altered teachers' views of teacher education" (p. 217). They noted the teacher turnover at the schools; as many as one-third of the teachers was new to each building, each year of the collaborative efforts. Additionally, the large sizes of the secondary schools proved problematic; the layouts of the schools, to some extent, made communication problems unavoidable. Furthermore, "skepticism" about university-faculty motives existed. Teachers at the schools "did not recognize theory as valuable or believe university faculty had unique and valuable contributions to make to pre-service teacher education" (p. 226). Their work with these secondary schools, at times, "got lost in the flood of school reform initiatives, each flowing from the principal's office" (p. 227). The power of departments in secondary schools was also addressed: "Relationships that teachers form with colleagues in their departments or in subject-matter organizations matter most in terms of professional development " (p. 228). Bullough and Kauchak (1997) agreed with Teitel (1993) that high program-participant constancy by university staff, teachers, and principals is a necessary condition for collaborative relationships to flourish and that program aims must be clarified early on.

Schoenfeld (2009) reiterated the critical nature of stakeholders in his account of professional development thar he spearheaded with the Berkley United School District (BUSD). At first, BUSD's superintendent refused to cooperate, and Schoenfeld noted that "school districts have been burned [by past professional development with university partnerships] and they are justifiably wary" (p. 199). However, Schoenfeld was not an "interloper" in this district; he was viewed as someone who could be trusted to have the kids' interests in mind. The professional development with BUSD incorporated "lesson study," and it was "assessment-driven" (teachers worked through mathematically rich assessment problems and then used them with their students).

Accordingly, it seemed that the professional development with BUSD would be easy to plan and to implement. However, Schoenfeld noted, "I had an agenda... my agenda and theirs were 2 different things" (p. 201). At the beginning of the first year, Schoenfeld allowed the teachers to share their "experiences, frustrations" (p. 202). During the program, which lasted five years, at times there existed confusion and havoc brought about by "levels of administrative chaos... layers and layers of bureaucracy... interwoven with a tradition of autonomy... distributed authority and lack of accountability" (Schoenfeld, 2009, p. 207). However, during year two, Schoenfeld began using student-assessment data to guide professional development, and he described year five as a *major turning point* because discussions about students' results on achievement test items became productive.

Despite the turbulence and the limitations he experienced, Schoenfeld noted, "It is possible to point out critical issues and set a number of productive activities in motion" (p. 209–210). He listed lessons learned that included: (1) spend time in teachers' classrooms; (2) build trust as a basis for professional interactions; (3) expect change over time; (4) realize that "complexity jumps by at least an order of magnitude if you operate at the systemic level"; and (5) find and nurture talent and connections (p. 216). Finally, Schoenfeld noted that "if I had to do it over again I would ask for direct access to decision-making groups [district level] as a matter of policy" (p. 216). Ultimately, Schoenfeld noted the importance of using student-assessment data to drive professional development and the critical nature of shared responsibility from stakeholders. His work with BUSD paralleled our work with a large urban school system.

Evaluation of Professional Development

Evaluation of professional development typically does not assess student learning (Loucks-Horsley and Matsumoto, 1999). Based on research conducted by the Education Commission of the States (ECS, 1997) at 16 school districts, "Most evaluations of professional development were focused on compliance rather than on effectiveness and did not evaluate the connection between the dollars spent, the programs purchased, and the results they obtained" (ECS, 1997, p. 7). If teacher learning is critical in helping instruction move beyond mechanistic implementation to maximize student learning (Loucks-Horsley and Matsumoto, 1999), then professional development should measure both teacher learning and student learning.

In fact, The Middle Grades Initiative of the National Staff Development Council conducted research and found that more than 90 percent of professional development programs had no measurement of student achievement (Killion, 1998). D'Ambrosio, Boone, and Harkness (2004) used survey data to compare and contrast students' perceptions of their teachers' practice and the teachers' perceptions of their own practice in the classroom; these data were then used to develop professional development experiences for the teachers that increased inquiry and discovery in the mathematics classroom. No measure was developed to find out whether or not the professional development program impacted students' beliefs about or achievement in mathematics.

As noted previously, using student-assessment data defined a *major turning point* when Schoenfeld worked with the BUSD. Guskey would, perhaps say, I could have told you so. If we *start* professional development planning with "the desired result—improved student outcomes" (Guskey, 2002, p. 45) results would prove to be more successful. Traditionally, we have not paid much attention to evaluation of professional development except at a surface level (Guskey, 2002). In fact, Guskey identified five "Critical Levels of Professional Development Evaluation" (see table 2.1).

Each level builds upon the preceding level and as each level is assessed, evaluators should look for evidence of areas of strength and/or weakness in professional development programs (Guskey, 2002, p. 49). *Level 1* is easiest to evaluate and focuses on whether or not participants liked the experience and whether or not their basic human needs were met (e.g., good food, comfortable room temperature); assessments at this level can help providers improve the design and delivery of the professional development. *Level 2* assessments attempt to answer questions about what participants' learned as a result of the professional development; measures at this level show that specific learning goals for teachers were met and, therefore, these indicators need to be established prior to the beginning of the professional development. *Level 3* focuses on organizational change and policies that either allow or hinder participants' to use what they learned from the professional development; unfortunately, problems at this level can cancel the gains made at the first two levels. *Level 4* information cannot be gathered during the professional development program because time must be allowed for participants to adapt the new ideas and

Table 2.1 Critical Levels of Professional Development

Level	Component	What Was Measured	Tool Used to Collect Data
1	Participants' Reactions	Participants' initial satisfaction—"happiness quotients"	Questionnaires
2	Participants' Learning	New knowledge and skills of the participants	Participants' reflections and portfolios; simulations; demostrations
3	Organization Support and Change	The organization's advocacy, support, accommodation, facilitation, and recognition	District and school records; structured interviews with participants and district or school administrators; et cetera.
4	Participants' Use of New Knowledge and Skills	Degree and quality of the implementation	Direct observations; video or audio tapes; structured interviews with participants; et cetera.
5	Student Learning Outcomes	Student learning outcomes: performance, achievement, dispositions, et cetera.	Student records; test data; questionnaires; structured interviews with students, parents, teachers, and/or administrators; et cetera.

Source: Guskey, 2000.

practices into their classrooms; a key tool for evaluation at this level is direct observation or videotape of teachers in their own classrooms. *Level 5* addresses whether or not professional development affected students as established in the program goals; multiple measures (e.g., achievement test scores, grades, dispositions surveys) should be used to assess this level. Key t o using Guskey's "Critical Levels of Professional Development Evaluation" is the notion that in planning *the order of these levels must be reversed*; start working backwards with *Level 5* by deciding on student outcomes that will be impacted.

Guskey (1998) noted,

> Tracking effectiveness at one level tells you nothing about impact at the next. Although success at an early level may be necessary for positive results at the next level, it is clearly not sufficient. This is why each level is important. Sadly, the bulk of professional development today is evaluated only at *Level 1,* if at all. Of the rest [other professional development programs], the majority are measured only at *Level 2* (p. 44).

In regards to professional development schools, Levine (2002) concurred with Guskey; student learning should define the mission, curricula, and the direction of research and inquiry for teacher candidates, school faculty, and university faculty. When student learning was the focus of planning, test scores of students in professional development schools in both Michigan and Texas met, exceeded expectations, and showed gains over time (Proctor, 1999; Pine, 2000). Unfortunately for us, we were not involved in the planning for the professional development program *MTL*, described in the next section. At times, we felt like pawns who were trying our best to work with teachers and not step on the toes of administration.

Our Work in MTL

Our *MTL* site team included the school district's curriculum coordinator, two mathematicians, two mathematics educators, and three graduate students. The *MTL* program-specific outcomes—created by the program development leadership group, not the site team—were to gain administrative support where the model for professional development focused on developing certain practices: (a) teachers continue to learn and do mathematics, (b) teachers reflect upon and refine their practice, and (c) teachers become resources to their colleagues and the profession (program documents). Ultimately, these efforts would impact student outcomes but the outcomes were not explicitly articulated or defined.

Year One

Our site team envisioned a professional development program specifically tailored to each of the three selected schools (X, Y, and Z). By helping teachers create individualized programs at each school, we anticipated that they would be actively involved, have vested interests in the programs, and find them worthwhile. Recall that this is what NCTM recommended in 1991 (*Professional Teaching Standards for School Mathematics*). In the fall of 2004, the site team attended department meetings and talked with teachers about the types of professional development that they thought would be most useful to them. We projected that after three or four months, with our support, each school would have designed and begun to implement its own professional development program. Additionally, after the creation of the professional development plans at each school, the site team intended to divide itself among the schools; the university professors would work with one school each, and the graduate assistants would help the professors as needed.

However, we had not anticipated the difficulty of getting the teachers to take charge of creating and designing their own professional development. During the first few months, the teachers typically talked about the challenges they faced in their classrooms and the strain that the No Child Left Behind Act of 2002 (Grobe and McCall, 2004) placed on them. As the months wore on and no professional development programs were designated, teacher attendance at the monthly meetings began to decline, a circumstance indicative of low levels of administrative support. At the end of the first year, all meetings at School X had stopped due to low attendance from the teachers; however, through increased pressure from the site team, teachers at Schools Y and Z met the last months of the school year and created plans for professional development that were targeted to begin the following school year. While plans were in place at each of these two schools, they were mainly the ideas of the site team rather than the teachers at those schools.

Year Two

In September 2005, Shelly was recruited to work with the five teachers at School Z. Schools X and Y were designated to other members of our site team; however, the professional development activities in both of those schools waned and eventually were discontinued. Only the work in School Z continued after about October.

Although teachers at School Z had identified "Japanese Lesson Study" as their professional development focus at the end of year one, we felt that this was not a plan the teachers were interested in. At the first meeting at School Z, we reopened the discussion about what focus the teachers would find most helpful. One teacher in attendance mentioned that students had difficulty in solving "simple" equations (e.g., $x + 17 = 56$ or $2x - 25 = 37$). We offered to introduce the teacher to Hands-On Equations™ (www.borenson.com), a method for teaching equation-solving using manipulatives within the context of a balance scale. Other teachers expressed interest, and a Hands-On Equations™ demonstration was planned for the next department meeting.

The eventual theme for the professional development at School Z was the Representation Standard, one of the five Process Standards, newly added to the *Principles and Standards for School Mathematics* document published by the NCTM (NCTM, 2000). This theme seemed to emerge from the Hands-On Equations™ demonstration at the second monthly department meeting at which we shared Bruner's (1966) research and explained that in order for students to fully grasp the symbolic representation and manipulation required to solve equations, they needed to first experience concrete and pictorial representations. Three of the five mathematics teachers at School Z were open to allowing us to model the use of Hands-On Equations™ and other lessons with concrete and pictorial representations in their classrooms. We spent one day each week at School Z and visited each of the five teachers' classrooms on a rotating basis. After a few weeks, three of the five teachers began to invite us into their rooms and asked us to co-teach or demonstrate other lessons with multiple representations. Typically, these lessons dealt with algebra topics. The other two teachers invited us to observe, but not to co-teach or demonstrate lessons.

Based on this show of interest, we planned a research study and obtained approval to investigate changes in teachers' understanding and use of representation over the course of the school year. Unfortunately, the Institutional Review Board approval was not finalized until April. Even though all five teachers agreed to participate it was nearly impossible to investigate "changes" over a span of two months (April and May of 2006). In an effort to document any change, teachers completed surveys and participated in interviews about their use of different forms of representation in lesson planning and implementation. We continued to visit School Z weekly, modeling or observing lessons. Monthly department meetings were held for the purpose of having group discussions about: how the modeled lessons had worked with students; readings sent to teachers about using different types of representation; and mathematics activities. In order to meet one of the program-specific outcomes—created by the program development leadership group—related to developing *MTL* ("teachers become resources to their colleagues and the profession"), beginning in January we encouraged the mathematics department chair to take over responsibility for the monthly meeting agenda. Unfortunately, in order for this to happen we had to send numerous email reminders; at some meetings in the spring there were no agendas created by the mathematics department chair.

Grant funds were used to provide teachers with books, which contained examples of lessons that incorporated the use of manipulatives such as geoboards, two-color counters, pattern blocks, and others. The site team used funds to lead two after-school "Representation Workshops" in April and May during which the teachers investigated mathematics problems, which incorporated concrete, pictorial, and symbolic representations. Additionally, funds were used to buy two classroom sets of Hands-On Equations™, pay for one teacher to attend a regional conference, and to cover the cost of a substitute teacher in order for another teacher to observe mathematics teachers at other schools in the district.

Because *MTL* was not funded for a third year, our initial feelings were that of failure and disappointment. After time passed we were able to analyze the program (see the Ratings section) and talk about and assess what had gone wrong. Subsequent to evaluating the program as a whole, we took a closer look at what, if any, effect the professional development had on the individual teachers. Through the process of looking more closely at the data, we were able to see that in fact there were successes.

The Ratings

In order to analyze the program, we decided to use the *OSPD* contained within the Ohio Department of Education's *Standards for Ohio Educators* (2007). This document was released about a year after our work at School Z ended. According to the standards-writers, the *OSPD* "define the characteristics of High Quality Professional Development (HQPD)" by the following six standards:

Standard 1—Continuous: purposeful, structured, and occurs over time
Standard 2—Data-Driven: informed by multiple sources of data
Standard 3—Collaborative: stakeholders share in the responsibility of outcomes
Standard 4—Varied: learning experiences accommodate individual teachers' knowledge and skills
Standard 5—Evaluated: assessed based on its short- and long-term impact on professional practice and achievement of all students
Standard 6—Results-Oriented: teachers acquire, enhance, or refine their skills and knowledge

We used these six standards and the "Elements" and "Indicators" for each standard to evaluate our work with School Z. For example, Standard 3 includes three Elements. One of these, "3.1, Professional development provides ongoing opportunities for educators to work together," contains three indicators:

1. Educators have the knowledge and skills needed to collaborate in teams successfully.
2. Collaboration is supported by creating opportunities for flexible scheduling of participants.
3. Participants are provided opportunities to meet regularly in collaborative teams to focus on improving practice and student achievement (p. 68).

We created a rating scale to evaluate our performance against each individual Standard 1–6 and its Elements and Indicators: Pass; C; or Fail. A designation of "Pass" indicated that our efforts addressed and met the characteristics of the standard for HQPD. A "C" designated that our efforts addressed some characteristics of the standard for HQPD but did not address or meet others. A "Fail" indicated that we were completely unsatisfied with our efforts and did not meet any of the characteristics of the HQPD standard. To summarize our ratings we created table 2.2. Please note that we found overlap in Indicators; for example, the fact that we had a weekly presence in the school and were working with individual teachers fit within Indicators for Standards 1, 3, 4, and 6.

Standard 1: Rating = C

HQPD includes multiple steps, "planning, implementation, reflection, evaluation, and revision" (p. 62). The participants should be involved in the planning. This was exactly what we envisioned would happen during year one of *MTL*. At School Z our decision to focus on the theme of the Representation Standard (NCTM, 2000) was guided by the teachers' consensus that their students had trouble solving equations; however, they did not have complete ownership about the exploration of how "representation" impacts students' learning because this process standard theme was suggested by us, the researchers.

HQPD participants must also be provided with time "to apply new ideas and to reflect on changes in their practice" (p. 62). We felt as though opportunities for this to happen abounded because we had a weekly presence. After our classroom visits we attempted to engage in reflective conversations about the teaching and learning; also, at the monthly meetings we endeavored to facilitate conversations about using multiple forms of representation to teach mathematical concepts and skills.

Table 2.2 Ohio Standards for Professional Development and Program Ratings

Summary of Effort Ratings Standard	Description	Our Rating
1	Continuous	C
2	Data-Driven	Fail
3	Collaborative	C
4	Varied	Pass
5	Evaluated	C
6	Results-Oriented	C

HQPD resources are "made available and allocated" (p. 62) so that the teachers can implement their new skills and knowledge. If they needed manipulatives or books or other supplies, we purchased these for the teachers through program funds.

Additionally, HQPD leaders also identify and make salient the goals of the program. In hindsight, we were not clear about the goals of *MTL*, so it is now obvious to us that the teachers did not know the goals.

Standard 2: Rating = Fail

The ultimate goal of HQPD is to increase student performance. "To ensure that educators perceive the value and relevance of professional development, educators must be involved in analyzing data, research, and best practices to determine the focus of the professional development" (p. 65). Because our focus on NCTM's Representation Standard resulted from a teacher comment about students not being able to solve "simple" algebraic equations like $4x + 8 = 20$, it was not data-driven in the form of students' standardized test scores.

When we gave the teachers readings about best practice and the use of representation, we hoped these would inform our conversations during the monthly meetings; however, during post-interviews we found that most of the teachers either did not read or did not remember the readings.

Standard 3: Rating = C

HQPD must provide "ongoing opportunities for educators to work together" (p. 68). This standard aligned with *MTL*'s goal of "teachers become resources to their colleagues." In addition to our monthly meetings, we met twice to investigate mathematical content during "Representation Workshops." Furthermore, two teachers collaborated to co-teach a lesson, which incorporated multiple representations.

HQPD should also be planned, delivered, and evaluated by a diverse team of educators. There were forums for teachers to share and discuss their classroom experiences with colleagues during monthly department meetings, to provide feedback on the activities during workshops, and to share their thoughts about what they found to be valuable or not during interviews with the site team at the end of the academic year. However, just as teachers were reluctant to participate in the creation of the professional development focus, they did not seem inclined to evaluate the overall program or suggest adjustments that could be made to make it more valuable. While the site team made every effort to create an atmosphere in which the teachers considered themselves as cocreators and coevaluators of *MTL*, we do not think this was realized.

Standard 4: Rating = Pass

HQPD must meet the needs of individuals and the group as a whole. Our site team worked with both individual teachers and the entire group of five teachers and tried to help teachers "refine or replace previous knowledge and skills" (p. 70).

HQPD experiences should be matched not only with the identified needs, but also with the knowledge and skills of the educators. During the discussions that followed the teachers' identification of solving equations as troublesome for their students, we discovered that the teachers were not familiar with NCTM's Representation Standard, the work of Bruner (1966), or teaching methods using manipulatives. Based upon this information, we tailored the *MTL* around these areas of best practice and varied our roles in individual teacher's classrooms depending on the level of comfort each of them showed with regard to teaching using multiple representations. In some classrooms, we first taught a lesson, modeling techniques, while in other classrooms we acted to support teachers when they used new manipulatives or representations. Post-interviews conducted six months after our work at School Z ended revealed that all but one teacher reported their continued use of multiple representations.

Standard 5: Rating = C

HQPD ought to be delivered with an evaluation plan in place to measure the impact of the experience on levels of: teacher participation, satisfaction, learning, and implementation; student learning; and school culture. Looking back, it is clear that we did not have a structured evaluation plan in place, one that was informed by the overall goals of the grant. Our evaluation plan was primarily guided by our research questions, investigating changes in teachers' understanding, and use of representation over the course of the school year.

In meetings with teachers, we asked for their assessment of the impact of lessons we modeled or assisted with on student learning. In a post-interview, one teacher talked about how students liked the activities and asked when we were coming back. A teacher who had fully implemented Hands-On Equations™ shared the belief that students had a much better conceptual understanding of equation solving because of using the balance representation. Another teacher told us that a lesson we modeled using two-color counters "hadn't worked" but later admitted that it had worked for at least one student. Teachers remained positive or seemed more positive about teaching methods when they saw an explicit impact on students' attitudes and/or learning.

Standard 6: Rating = C

HQPD must have an impact on professional practice by enabling teachers to increase their knowledge of content and pedagogy and supporting them in putting what they have learned into practice. We focused on research-based best practices and provided numerous examples of applications to the classroom; we were also available to support teachers each week. Some of the positive changes we observed in teachers, listed in the next section, were changes in classroom practice.

Small Successes

Because *MTL* was not funded for a third year, our initial feelings were that of failure and disappointment. After time passed we were able to evaluate our efforts. Moreover, through the process of looking more closely at the data, despite the turbulence we felt when we worked with the program, we were able to see that in fact there were successes. As mentioned previously, subsequent to evaluating the program as a whole, we took a closer look at what effect, if any, the *MTL* program had on the individual teachers. Based on this analysis we created table 2.3.

Although we viewed these as "small successes," we were left wondering if this language did justice to our efforts. Perhaps "small successes" were actually evidence of greater shifts in teachers' thoughts about teaching and learning mathematics than we can confirm with the data we collected (Teachers 4 and 5—table 2.3). Or, maybe the "small" can also refer to the fact that we ultimately only worked closely with five teachers. Birman et al. (2000) emphasized the notion that districts should provide HQPD by either serving fewer teachers or investing more resources. We served fewer teachers with more resources by default.

During monthly department meetings and the two "Representation Workshops" we attempted to provide opportunities for collegial interaction and reflection but found it hard for teachers to move away from talking about their frustrations. For example, even though Teacher 4 became less vocal about students' weak skills and poor work habits (see table 2.3) in May she said,

> At this school, where 46% only of students pass the Ohio Graduation Test and it's easiest test that I saw... If it's class like only 2 positive grades, you know, something like 22 students and only 2 positive grades, you go to concrete. You know, abstract doesn't work at all.... even in 8th grade I didn't

use this stuff. We use it in 4th, 5th grade.... At least at this school they need it [concrete and pictorial representations].

Teacher 4 said, she was comparing students' weak mathematical skills at School Z with her own mathematical skills and, therefore, students needed to use concrete and pictorial representations. In other words, mathematics is about abstract concepts, not about making sense of those concepts with

Table 2.3 Evidence of Success in Project

Small Successes with Individual Teachers	Evidence of Success
1	Allowed site team members to demonstrate lessons to students; co-taught a "representations" lesson with Teacher 2; copresented at the Ohio Council of Teachers of Mathematics with one of the site team members; after the program ended, mentored three student teachers and helped them create lessons which incorporated multiple representations
2	Used Hands-On Equations™ extensively; co-taught a "representations" lesson with Teacher 1; took a "calculus for teachers" course; became more vocal and challenged Teacher 5 at monthly department meetings; after the program ended, continued to stay in touch with one site team member for at least three years
3	Expliitly "liked" the "Representation Workshops"; became more open to let site team members observe some lessons; taught a lesson, using a concrete representation for solving a system of three equations in three unknowns (gleaned from The Mathematics Teacher an NCTM journal for high school teachers)
4	Become less vocal about students' weak skills, poor work habits, and so on; began to demonstrate concepts using concrete models but did not allow students to use concrete models (although this may seem trivial, it seemed a major breakthrough to us!)
5	Allowed site team members to demonstrate lessons to students; allowed site team complete access to the classroom; came to some meetings in year two (did not attend any meetings in year one); seemed to value participating in the research aspect of *MTL*; admitted that a two-color counter lesson related to integer operations "helped one student" (although this may seem trivial, it seemed a major breakthrough to us!)

concrete and pictorial representations. However, we must emphasize the fact that the *students* did *not* use concrete or pictorial representations in her class. She said, "*I* do it a lot if I have chance to put a picture or graph [*sic*]." At the end of the interview, Teacher 4 noted, "So, thank you. I am very thankful for this program.... It's not been easy to work with us, but thank you very much."

Teacher 2, in May, said, "I now understand that kids need a concrete way to see concepts. They also need pictorial." However, she also noted that, "it's hard to find the activities and I want to make sure it [the activity] relates to the concepts." Additionally, she talked about how her students struggled with solving equations and that Hands-On Equations™ helped them, "mentally, because they pictured moving the pawns to balance the equations." She also thought students "really understand slope" because of the Bottle Filling Activity, the Bridges and Pennies Activity, and other activities that Shelly modeled in her classroom, "and, students were better at finding the nth term because of activities like The Bees in the Trees." Typically, Shelly would model the activity in one class period, and then Teacher 2 would use the activity in her other classes.

> Teacher 5 talked about student learning. She said,
>
> Shelly came last year and did some [activities] with the kids, in all fairness, there were a couple that had terrible self-esteem, or confidence and [the students] did a real good job and focused on her [Shelly]. Um, and, learned better because of it. A couple. Um, my classes are too large.

Although she admitted that the activities (which focused on letting students make sense of mathematics through the use of concrete representations) that Shelly used helped a couple of students, Teacher 5 nevertheless needed to vent about her class sizes. She also noted, "Very few of them [students] come in knowing something. I probably have five." It was hard to have conversations with Teacher 5 in which the focus was not on her frustrations. She saw little value in collegial interaction and conversations about teaching and learning in monthly meetings because she thought most of her students came to her class not "knowing something" about mathematics.

However, as noted in figure 2.3, Teachers 1 and 2 collaboratively planned and co-taught a lesson in which their students used concrete and iconic representations to solve geometry problems. The collaboration was especially important to Teacher 2. She sought advice from Teacher 1, and that led to the co-taught lesson development and implementation. In fact, research indicates that collegial interaction does not automatically occur when teachers are placed in a room together (Taylor, 2004). We concur, especially when teachers have different beliefs about why their students are not successful and simply want to place blame.

In December of the year after *MTL* was terminated, we conducted follow-up interviews with Teachers 1, 2, 3, and 5. Teacher 4 was no longer teaching full time in the district. The interviewer asked, "In what ways have you changed or not changed your lesson planning based on the use of multiple representations?" Their responses are listed below in table 2.4. Recall that Teitel reported that his work with secondary schools sometimes "got lost in the flood of school reform initiatives, each flowing from the principal's office" (Teitel, 1993, p. 227). This was true of our work at School Z except most initiatives were mandated by the district rather than the school principal. Mathematics teachers were pulled in many different directions for other types of professional development during the time we worked at School Z.

Additionally, district personnel and even our site team personnel changed over the two years we worked with the *MTL* program. During year two, one of the mathematicians knew he was leaving his position for a promotion at another university. Also, at about midway through the second year of the program, the school district's curriculum coordinator—a member of the local site team—learned that her job position was to be cut. The graduate students who worked with the program changed, and Shelly was not brought on board until year two. It is clear that although efforts were made to make the program continuous and seamless, it was not.

Table 2.4 Multiple Representation Responses from Participants

Follow-Up Interview Teacher #	Response
1	"I did a lot of teaching with representation to begin," so there have not been a lot of changes in my teaching. However, in the engineering class, "there's a lot more [labs, hands-on experiences] in that."
2	Last year I used Hands-On Equations™, and "I actually used it again this year." It depends on the topic and "if I think it's productive." Probability lends itself to using multiple representations.
3	I used a lesson in which students created models for solutions to systems of three equations in three unknowns, and "they understood it [why the solutions are inconsistent or consistent] better."
5	"I have not. I do not use Hands-On Equations™."

The Importance of Up-Front Evaluation Planning

According to the *OSPD* (2007) Standard 5—Evaluated—professional development should be delivered with an evaluation plan in place, in order to measure the impact of the experience on levels of: teacher participation, satisfaction, learning, and implementation; student learning; and school culture. These align with Guskey's five "Critical Levels of Professional Development Evaluation" (2000). However, the *OSPD* standards-writers did not emphasize the importance of beginning planning by creating an evaluation plan up front. Additionally, they did not emphasize the importance of the *order* of the evaluation planning.

Recall Guskey's (2002) view that when planning for professional development *the order of the levels must be reversed.* That type of planning might follow a course similar to this one:

- Begin planning by assessing student-achievement data, and use these data to plan for student learning outcomes (*Level 5*);
- Use these student learning outcomes to plan for how the professional development will be implemented in order to impact teachers' use of new knowledge and skills (*Level 4*);
- Move from the planning for teachers' new knowledge and skills to plan for how to gain school and/or district support for the professional development (*Level 3*);
- After this support has been the focus of planning, decide what kinds of activities will help foster teachers' acquisition of new knowledge and skills (*Level 2*);
- Finally, create strategies that take into account teachers' "happiness quotients" or initial satisfaction with the professional development (*Level 1*).

It seems as though the *OSPD* Standard 2—Data-Driven—is most critical to Guskey's *Level 5.* According to the *OSPD,* the ultimate goal of HQPD is to increase student performance, and data that reveal student performance should be analyzed. "To ensure that educators perceive the value and relevance of professional development, educators must be involved in analyzing data, research, and best practices to determine the focus of the professional development" (OSPD, 2007, p. 65).

Remember the focus of the professional development at School Z was on the use of representation (NCTM) that was determined by us, and by default, because after more than a year the teachers themselves did not agree on a focus. Therefore, our professional development activity was not data-driven in the form of students' standardized test scores or other measures from students. (One could perhaps argue that the focus was informed by teachers' use of qualitative data or at least by one teacher's remark that the students struggled to solve "simple equations"). Interestingly, we rated our efforts within the *OSPD* Standard 2 as "Fail." We now wonder if our efforts would have been more successful had the professional development leadership group (coordinating the program from outside the three regional sites in the Midwest, Northwest, and South) insisted or encouraged local site teams to use student data to drive the focus of the professional development efforts at each school. More specifically, what might have happened had the program started with articulated goals, focused on student achievement and was data-driven (*Level 5*), and with an evaluation process in place that worked backward through the other four levels (as per Guskey's recommendations)?

Looking back, it is clear that we did not have a structured evaluation plan in place, one that was informed by the overall goals of the grant. Our evaluation plan was primarily guided by our own research questions, investigating changes in teachers' understanding, and use of representation over the course of the school year.

Conclusion

Despite all of the ways that the *MTL* program was not successful, there were small successes with individual teachers. In hindsight, we feel that the biggest obstacle we faced was that we were never quite clear about the program goals. Up-front evaluation planning did not occur in conjunction with the program development leadership group, the local site team, and the teachers. Therefore, without up-front evaluation planning and articulated goals the program was, perhaps, doomed to fail or at least the chances of success were diminished. If we had used student-achievement data as we defined and articulated the goals, and then worked through Guskey's (2000) "Critical Levels of Professional Development Planning" might we have had success at School Z? Or perhaps, a better question: Might we have had "more" or "bigger" successes with teachers at all of the schools?

Chapter 3

Science Cooperatives in Missouri and Iowa (Science Co-op): Addressing the Needs of Small Rural School Districts of *Science Literacy for All*

Larry D. Yore, James A. Shymansky, Leonard A. Annetta, and Susan Everett

There is a need for research involving professional development (PD) and implementation of the *National Science Education Standards* (NSES; National Research Council [NRC], 1996) that systematically considers the theoretical and practical demands placed on decision makers, teachers, and researchers. However, few projects go beyond a single site to multiple schools, school districts, or state systems. Single classroom, school, or school district efforts fail to recognize the complexity of education systems and subsystems, scaling (capacity building), planned change, and the logistical, financial, and time demands involved in large-scale, multidistrict, multiagency efforts. The NSES content, teaching, and program standards focus on conceptual understanding, inquiry teaching practices, learning opportunities, and resources leading to science literacy and on fuller participation in the public debate about socioscientific issues. This chapter focuses on a multidistrict PD project that considered inquiry teaching and resources and its influence on reform-based classroom practice and student performance on high-stakes tests during a context of changing priorities, funding, and participants.

Background

The National Science Foundation (NSF) has funded various systemic change efforts such as state systemic initiatives (SSI, 1991–2001), urban systemic initiatives (USI, 1993–2006), and local systemic change (LSC, 1995–2006) projects (Kahle, 2007). These PD initiatives were intended to enhance teachers' science content knowledge (CK) and pedagogical knowledge (PK) related to specific hands-on curriculum materials (PCK) on the premise that knowledgeable teachers using high-quality curriculum materials would translate into improved classroom practice and student performance. The LSC initiatives were based on the need for highly qualified teachers of science and an inferred theory for action: (1) high-quality PD and (2) the availability and utilization of high-quality instructional resources would lead to (3) improved inquiry-based teaching practices translating into (4) improved student performance (Banilower et al., 2006). This *theory* appeared to be based on reasonable evidence and strong linkages amongst the first three factors but on little evidence linking them to factor four—student learning. Surprisingly, it was not until the final cycle (2001–2006) of LSC funding that projects were required to document impact on student achievement—and then, this requirement was applied retroactively to previously funded projects. Only four LSC studies of science learning and teaching have published their achievement results widely (Czerniak et al., 2005; Geier et al., 2008; Revak and Kuerbis, 2008; Johnson, Fargo, and Kahle, 2010)—beside the project that is the focus of this chapter. Other projects provided documentation plans and evidence to Horizon Research Inc. (HRI) for the LSC report (http://www.horizon-research.com/reports/), but very few reported high-stakes test achievement data—the normative value of most federal, state, and local policy makers and decision makers.

Unlike most LSC projects that focused on tightly located urban and suburban systems, the Science Cooperatives in Missouri and Iowa (Science Co-op) project focused on the elementary/middle school science programs of underserved rural and small school districts dispersed across two Midwestern states and 40,000 square miles. These recently consolidated school districts were geographically remote (separated by 10–15 miles of poorly maintained and icy country roads and by furious independence residual from preconsolidation time). Many schools had one or maybe two teachers teaching at the same grade level without resident content specialists and with limited financial resources, which required the teachers to rely on regional education and colleges/universities for piecemeal PD and implementation support of any new curriculum and instruction directives. The geographic and staffing characteristics of these districts and schools minimized collegial networking and leadership in science education. However, the lingering independence and competitiveness from historical athletic competitions and political negotiations among schools and school districts within specific regions appeared to be limiting effects for collaboration and mutual gains. The Science Co-op project assumed from the outset that success would be based on good engineering in designing solutions that addressed local problems, priorities, constraints, available resources, and isolation. The project's title reflects the basic metaphor for design and problem solving associated with rural America—the farm cooperative used in rural areas to address economic, technological, and political demands placed on small farmers and to share resources (equipment, expertise, etc.) to achieve common goals and financial success. Science Co-op utilized the same ingenuity, creativity, problem solving, and determination to address long-term PD and systemic change, interactive-constructive learning, science literacy, cross-curricular connections, and inquiry teaching.

Professional Development and Systemic Change

Banilower et al. (2006) synthesis of core evaluations of LSC projects revealed major successes and challenges across the nearly 100 K–12 mathematics, science, and technology education programs. They reported successes in three components of the LSC theory for action: quality of PD and support context, use of inquiry materials, and impact on teachers and their classroom practices. Challenges involved preparing and supporting PD providers, teachers' content needs, reaching the targeted teachers, day-to-day support of teachers, and the engagement of principals.

Banilower, Heck, and Weiss (2007) explored the relationships amongst teachers' self-reported information (i.e., attitudes toward NSES-based teaching, perceptions of pedagogical and content preparation, use of traditional and investigative teaching practices, use of practices to promote investigative culture, frequency of NSF-funded instructional resources, science instruction time) and hours of PD in 42 K–8 LSC projects. Analyses revealed complex models, significant nonlinear relationships, and small positive effect sizes among the amount of PD and these dimensions. Earlier, Supovitz and Turner (2000) demonstrated significant nonlinear, direct relationships between teachers' hours of PD and self-reported use of inquiry-based teaching practices and an investigative culture in the classroom using similar data sources and analysis techniques. They identified 80 hours of PD as the critical point for both relationships in which PD produces significant differences.

Other researchers have explored the effects of PD on teachers' perceptions about science teaching, nature of science, and classroom practices. Khourey-Bowers and Simonis (2004) investigated middle school teachers over a ten-year period and found significant pretest-posttest gains in three of the four cohorts for self-efficacy and outcome expectancy. Respondents reported greater comfort with the science content, assessment strategies, classroom organization, and teaching strategies leading to useable PCK and beliefs about successful performance. Akerson and Hanuscin (2007) reported that three elementary teachers demonstrated changes in their views of the nature of science, from traditional absolutist toward more modern tentative views, after a two-year PD program consisting of monthly half-day workshops. They also found that these teachers had more positive dispositions toward inquiry-oriented teaching, and their students expressed more modernist views of science. Johnson (2006) found that even within effective, supportive two-year PD programs teachers still encountered technical, cultural, and political barriers related to time, resources, and administrative support for implementing the desired instructional practices. She (2007b) found that classroom observation changes of a stratified subsample ($n = 6$) of teachers from two districts indicated that two teachers reached the desired level of implementation while four teachers demonstrated improvement to elements of the desired level. The results indicated an effect related to positive, prolonged, and committed efforts and support of one school/district compared to the other.

Promoting Science Learning

The committees on Developments in the Science of Learning and on Science Learning (NRC, 2000, 2007) suggested that much science education research has been based on outmoded views of learning and curricular resources that do not align with NSES goals related to *Science Literacy for All* students. The NRC (2000) stated, "Students often have limited opportunities to understand or make sense of topics because many curricula have emphasized memory rather than understanding" (p. 8–9). This report stressed three key principles: (a) that people come to learning with prior conceptions that must be engaged or challenged; (b) enhanced competence requires prior foundational knowledge; and (c) learning requires metacognition to be aware, monitor, and control meaning making. However, learning appears to advance and regress, producing a sawtooth pattern rather than a smooth curve (Shymansky et al., 1997).

The interactive-constructivist teaching model promoted in Science Co-op is a middle-of-the-road interpretation of constructivism. Interactive-constructivist teaching recognizes a specific worldview of thinking, the epistemological and ontological nature of science, the locus of mental activity in the learner, the sociocultural aspects of the classroom, the multiple purposes of language, and the realities of public education and schools (Henriques, 1997). This interpretation of science teaching assumes a hybrid view of the thinking that stresses the importance of interactions with the physical world and the sociocultural context in which interpretations of these experiences will reflect the learner's lived experience and cultural beliefs.

Several NSF-sponsored science programs partially address the interactive and constructive aspects of effective learning and doable instruction. The three most-used inquiry kit programs—Full Option Science System (FOSS; Lawrence Hall of Science, 2003), InSights (Education Development Center Inc., 1997), and Science and Technology Concepts (STC; National Science Resources Center, 2009)—utilize a variation of a learning cycle through which the varied experiences and instruction provide a basic platform to address cross-curricular connections to language arts, mathematics, social studies, and other disciplines and to elaborate the modified learning cycle (engage, explore, consolidate, and assess) teaching strategy (Henriques, 1997).

Science Literacy

Science Literacy for All is a long-promoted, but ill-defined, general expectation (Hurd, 1958) with international cachet (McEneaney, 2003). Analysis of *Science for All Americans* (American Association for the Advancement of Science [AAAS], 1990, 1993) and the NSES (NRC, 1996) reforms (Ford, Yore, and Anthony, 1997) identified interacting senses of science literacy in which the fundamental sense subsumes abilities, emotional dispositions, critical thinking, and information communication technologies (ICT) as well as language and mathematics—while the derived sense subsumes the content goals regarding understanding the big ideas of science, the nature of science, scientific inquiry, technological design, and the relationships among science, technology, society, and environment (STSE; Yore, Pimm, and Tuan, 2007). There is cognitive symbiosis between these senses and amongst the components within both senses leading to fuller participation in the public debate about STSE issues resulting in informed decisions and sustainable actions (Yore, 2009).

Science Inquiry Teaching

The NSES teaching standard emphasized inquiry approaches along a structured-to-open continuum: full, partial, or inquiry-like forms (NRC, 1996).

An analysis of the related pedagogical, cognitive, and functional demands of each approach indicates that idealized open inquiry may be too optimistic for most teachers and classrooms (Anderson, 2002). Johnson (2007a) stated, "Inquiry is a luxury, rather than a necessity; many teachers who use it periodically consider it to be in addition to the regular teaching of science, and oftentimes it is used as a reward for students after covering the required material." (p. 133). Yore et al. (2007) outlined real and imagined barriers to inquiry teaching for elementary and middle schools as teachers' background beliefs (goals of science education, nature of science, science literacy, science learning, science teaching), school characteristics (changing priorities, competing goals, external policies, organization, etc.), classroom climate (time, space, resources, classroom management, storage, etc.), and learner attributes (cognitive development, social behavior, background knowledge, etc.). They stated:

> In summary, teachers are often overwhelmed with the difficult task of implementing the more interactive and unpredictable teaching methods associated with inquiry and constructivism. Implementing this type of learning involves sophisticated integration of pedagogical skills and deep content. Learning and understanding do not come to students simply by the doing of activities (p. 64).

Johnson (2007a) stated, "It is clear that teachers require a great amount of support in order to teach science effectively, including the use of inquiry, cooperative groups, and classroom discourse" (p. 133).

Descriptive Context

The Science Co-op LSC project was based on several smaller teacher-enhancement projects that addressed the NSES, inquiry teaching, and authentic assessment. The most influential of those projects was Science: Parents, Activities, and Literature (Science PALs, 1993–1996).

The Science PALs Project

The Science PALs project was a teacher-enhancement project, funded by the NSF and the Howard Hughes Medical Institute, in which K–6 teachers and grades 7–12 science teachers participated in summer workshops and school-year PD activities. In addition to the teachers, approximately 3,400 parents participated in training sessions designed to integrate them into the

K–6 science instruction. Across the four years of Science PALs, teachers received an average of 110 hours of PD designed to get them to teach more inquiry science, integrate other disciplines, and enhance their PCK.

The project began with 16 elementary school teachers (most taught all subject areas in the school curriculum) who were designated as science advocates—one from each elementary school in the district. The science advocates attended a problem-centered summer workshop focused on the science kits scheduled for classroom use. The workshop matched university and secondary science teachers (partners) with small groups of advocates to explore science CK and PCK in specific units that the group selected, to adapt the kits, to promote interactive-constructivist teaching strategies, and to incorporate the NSES, students' ideas, children's literature, and parental involvement.

The project's cascading leadership design involved a progression from teacher as learner to teacher as leader responsible for teaching colleagues about the kits and strategies with parents. The summer workshop with follow-up in service was repeated in subsequent years with approximately 40 teachers in the second year, 80 the third, and 140 the fourth. PD focused on authentic problems of curriculum adaptation and on social interactions and private reflections to get the teachers to rethink their ideas. Teachers then used the same instructional cycle (engage, explore, consolidate, and assess) to challenge their students' ideas and to promote conceptual growth and change.

Science PALs enhanced classroom practice, students' science achievement and perceptions of science, and parental perceptions of the science program (Shymansky, Yore, and Hand, 2000; Shymansky, Yore, and Anderson, 2004; Yore, Anderson, and Shymansky, 2005). However, the project's legacy is the fact that the school district continued the PD and instructional approach after the external funding ended and put in place a science materials distribution center.

The Science Co-op LSC Project

The Science Co-op LSC project considered the system and subsystems involved in a five-year, multiple jurisdiction, multiple school district project and all factors and relationships in the four-part LSC theory of action (PD, resources, classroom practices, and student learning) from the proposal planning stage onward. We realized from our experience with Science PALs that the link between teaching and learning was tentative and the details about appropriate curriculum, assessment and materials management, planning and collaboration time for teachers, and supportive administrators, parents, and communities were critical barriers to science inquiry. Furthermore, we came to realize the complexity of long-term projects embedded in dynamic situations of changing leadership, staff, budget, priorities, competing goals, communities, and students as well as similar changes in the NSF program office and LSC requirements. This can be implied in Science Co-op: 2 state education agencies, 36 school districts, 90 elementary and middle schools, 1,100 classrooms, 1,250–1,450 teachers, and approximately 30,000 students over 40,000 square miles. These numbers partially reflect the enormity and complexity of this scaled project, but additional insights come from the fact that Iowa did not have an official statewide science curriculum and assessment, while Missouri had a state science curriculum and Missouri Assessment Program–Science (MAPS). Historically, Iowa ranks amongst the leaders in literacy performance and science achievement on the Iowa Test of Basic Skills (ITBS), while Missouri ranks below average in literacy and science achievement.

Science Co-op incorporated a cascading leadership model to develop local leaders and to manage and facilitate instructional changes in K–6 science, which evolved from project-centered leadership to development of a local advocate in each school and a local management team in each school district. The advocates and management teams gradually took over planning, facilitation, and delivery responsibilities in the third year that culminated in majority control and responsibility in the final year. The desired instructional changes involved moving toward an interactive-constructivist learning cycle teaching approach, adapting NSF-funded curriculum materials (e.g., FOSS, STC, InSights, combinations of kits and textbooks, local units, etc.), and developing local curricular supplements, resource people, and authentic assessment strategies.

Methods

A series of mixed-methods studies was used to document, establish, and describe the types and quality of PD enacted across the implementation of the cascading leadership model, the availability and utilization of inquiry resources, the enhancement of classroom practices, and the influences on student achievement. Both quantitative and qualitative data were collected and analyzed to explore these research foci. Quantitative data, as determined by the LSC-project evaluator (HRI) or normative sources of state and local authorities were used to establish relationships where appropriate; qualitative data were used to provide a rich description of the patterns and to identify potential influences. Both quantitative and qualitative data were selected to influence policy and decisions about science education at the state, district, school, and classroom levels; therefore, these data needed to be understandable and valued by the specific policy and decision makers (Shelley, 2009).

Data Collection and Analysis

Data were collected annually or biannually on the PD experiences, instructional resources (NSF-sponsored modules, science topics, and weekly lessons and instructional time), classroom practices, and student achievement. Random samples of PD activities (5–8), teacher questionnaires (300), teacher interviews (10), classroom observations (16), surveys of district resources (36), and highly valued grades 3 and 6 achievement measures were used. The observation instruments, questionnaires, and interview protocols were developed by HRI; observers were certified at annual training and calibration workshops. Science achievement was measured by the high-stakes tests (ITBS and MAPS) for participating school districts and a matched set of comparison school districts.

Data analysis and interpretation was complex and problematic because of ethical and technical considerations. Any instruction, resources, or achievement relationship identified in small rural districts could be associated with a specific school or teacher. Furthermore, the Science Co-op project was a systemic change study, which used school districts as the unit of interest. Therefore, all data were consolidated to the project level or reported as unidentified school districts to ensure anonymity. Other issues considered preestablished criteria by NSF or the original proposal. NSF established 130 (>129) hours per participant as the fully funded PD level. All districts and schools were expected to make available two inquiry-based units in each K–6 grade and to encourage all K–6 teachers to participate. Quality of PD was assessed by selecting five or more half-day or day-long activities annually and applying a HRI protocol. Quality of classroom practice was tracked by observing random samples of 16 teachers on a biannual basis and annual interviews of 10 teachers using HRI protocols. Each protocol was criterion referenced and used a hybrid analytical and holistic rating rubric in which specific dimensions and overall capsule ratings were required. The capsule rating was not a numerical sum or average of the analytical dimensions. All quantitative procedures provided ordinal or interval data receptive to statistical analysis. Observers' field notes, teacher interviews, summary observation reports, and external evaluator and director's annual reports were read and coded with common themes identified using constant comparison. Themes found in early information sources were verified or revised based on the interpretations of later information sources. Themes were used to develop assertions that were supported by illustrative quotes and information from these sources as evidence.

Results

The results are reported systematically to address the factors in the LSC theory for action: PD, resources, classroom practices, and student achievement. Assertions are in boldface, followed by quotations where available in italics, and descriptive and numerical information in normal font as evidence and elaborations of the assertions.

Professional Development

Documentation and reflections on the general and specific PD aspects revealed that the Science Co-op model was based on an annual cycle, which allowed for sustained support of science teaching during the school year, repeated visitation of problematic issues, and variation of the structure and content across the regional cooperatives and duration of the project. The participating teachers began with a one-week summer workshop to gain experience with adapting a self-selected FOSS, STC, or InSights kit in which teachers developed a teachers' resource book (TRB) that contained explicit connections to state or local benchmarks/learning outcomes, related misconception literature, modified learning cycle, cross-curriculum activities, additional science inquiries, local resource people, children's literature and trade books, and assessment techniques. The workshop contained sessions on authentic assessment, integration of reading/language arts and mathematics with science; the elements of inquiry (engage, explore, consolidate, and assess); and science content. During the school year, PD continued during monthly meetings at the local school district level and interactive televised (ITV) seminars that focused on developing science content or PK.

Face-to-Face PD Activities

ASSERTION #1: The PD was high quality across variations in persons providing the opportunities, focus topics, participants, and location of the experiences.

The year-one (Y1) summer workshop was held in a central location and involved one elementary teacher (advocate) from each K–6 school and one secondary science teacher (partner) from each district. Since the advocates and partners would become the local leaders, they participated in workshop activities similar to the ones they would lead to gain knowledge and confidence in mentoring their colleagues. They also participated in workshops solely focused on leadership, curriculum planning, and PD delivery. Advocates developed a TRB based on a science kit that included adaptations for state standards, local situation, and additions such as children's ideas, cross-curricular activities, and local resources and experts.

The summer workshops for Y2–Y5 were held at up to seven regional sites. The project staff worked with the advocates and partners at project-wide meetings to plan and prepare a similar workshop template rather than having distinct workshops in each region. This constraint and the use of external experts were reduced in Y3–Y5 to accommodate and address local needs, transfer responsibility (cascading leadership), and utilize local expertise in the specific district or cooperative. Each workshop was led by a project staff member, a science consultant who led the hands-on science kit work, a local school district facilitator, and advocates and partners who provided classroom expertise with specified kits. Some regional workshops were large with many school districts, thus the number of advocates (3–20), partners (2–7), participating teachers (25–84), and target science kits (6–18) was greater than some of the smaller co-ops. The variations meant that advocates, partners, and consultants had to distribute their expertise and time to different numbers of teachers (ratio 1:2.6–13.1) and kits (ratio 1:1–6.5). There did not appear to be any magic ratios amongst advocates, consultants, kits, and participating teachers, since most working groups had a rich distribution of background knowledge, classroom practice, and informal experiences with various kits and parts of the integrated science approach encouraged in the project.

PD sessions (five–eight half to full day) were observed and rated using the HRI PD observation protocol (http://www.horizon-research.com/LSC/manual/#2) that addressed (a) several demographic, focus, and contextual issues; (b) six analytical dimensions (design, implementation, content, pedagogy/instructional materials, leadership, and culture); and (c) a single holistic consideration. Each analytical dimension consisted of several key indicators and a synthesis rating in comparison to best practice in PD. Design focused on planning and organization: goals, tasks, roles, collaborations, and interactions; instructional strategies, activities, classroom resources and practices, and time and structure for constructing understanding, reflecting on these ideas, and sharing their insights with other participants. Implementation focused on the formal presentation: facilitator's contributions, modeling, background experience, expertise, and management style; pace of the session; and modeling of assessment. Content addressed the academic level, accuracy, and appropriateness of the science: representation of the nature of science, elements of the scientific enterprise, connections to real-world applications and other areas of science, engagement of participants, and extent of sense making. Pedagogy/instructional materials addressed the depth and breadth of attention to student thinking and learning: classroom strategies, instructional materials; facilitator's display of pedagogical understanding, participants' engagement, and sense making about classroom practices. Leadership focused on lead teachers and principals and addressed effective staff development: mentoring/coaching of peers, reform advocacy, facilitator's understanding of leadership, participants' engagement, and consideration of the content. Culture addressed encouragement and value of active participation, climate of respect, collegial working relationships, participants' directed knowledge building and sharing, academic rigor, constructive criticism, and debate. Individual dimensions were rated on a five-point ordinal scale (1 = not at all reflective...5 = extremely reflective of best practice) while the independent overall capsule rating (not a numeric composite of the dimension ratings) was rated on an eight-point ordinal scale (1 = ineffective passive learning highly unlikely...8 = exemplar active learning highly likely to enhance participant's capacity to provide high-quality instruction).

Table 3.1 displays the means and ranges of ratings for the PD activities formally evaluated across the project. The series of observations captures the cascading leadership (a fundamental design principle of Science Co-op) that involved decentralization of the PD from the central project staff and external consultants (Y1) to advocates, partners, and regional project staff (Y3) and finally to the advocates, partners, and local leadership teams (Y5).

Table 3.1 Descriptive Statistics for Professional Development Observations

Category	Year One (n = 5)	Year Three (n = 8)	Year Five (n = 8)
Design (1–5)	4.0 (3–5)	3.9 (3–4)	4.9 (4–5)
Implementation (1–5)	4.2 (3–5)	4.0 (3–5)	4.9 (4–5)
Content (1–5)	4.0 (3–5)	3.7 (3–4)	4.9 (4–5)
Instructional Materials (1–5)	4.0 (3–5)	3.6 (3–4)	4.9 (4–5)
Leadership (1–5)	NA	3.7 (3–4)	4.5 (3–5)
Culture (1–5)	4.2 (4–5)	4.9 (4–5)	4.9 (4–5)
Capsule Overview (1–8)	6.0 (5–8)	6.5 (6–7)	7.9 (7–8)

Mean ratings indicate PD quality (somewhat to extremely reflective of best practice) across the six dimensions and high-quality capsule assessment of the overall PD experience (highly effective, accomplished effective, or exemplary) for the duration of the project. PD quality was maintained as the responsibility and delivery was transferred from external experts and project staff to the regional and local leadership team. The somewhat lower ratings in Y3 for some dimensions as compared to the baseline (Y1) were a result of the initial transfer of responsibility. The culture category ratings and observers' comments in the HRI reports indicate the project's success in establishing a cooperative and supportive environment. The Y5 mean ratings returned to very high quality as the transfer to local leadership teams was finalized. The variability illustrated by the range across the observations was small or decreased for individual categories, indicating very good quality control across PD activities and providers for the decentralization associated with cascading leadership.

Distance Education PD Activities

ASSERTION #2: Participants preferred live ITV to recorded synchronous and asynchronous delivery modes but valued the flexibility of time and space afforded by an asynchronous mode.

Distance education was employed with an original plan to deliver content through live interactive videoconferencing. As technology evolved, two new ITV methods were used to flex delivery and participation and maximize use of external experts: synchronous video and asynchronous web-based modes. The need for the asynchronous mode was due to increased pressure from university and videoconferencing network administrations and the desire to increase participants' responsibility and flexibility for their professional learning.

The live sessions were conducted from an origination site (any node in the Iowa or Missouri ITV networks) from which the expert presenter and the session facilitator broadcast to up to eight remote sites. At the remote sites, teachers met in small groups in ITV rooms that contained television monitors on which they could see and hear in real time the presenter or facilitator and participating teachers at the other sites. A camera mounted on the back wall of the origination site captured the presenter and facilitator, and a camera mounted on the front wall of the remote sites captured the participants. Control panels at the originating site allowed the presenter or facilitator to navigate amongst the active sites or respond to requests from specific sites. Audio was captured through microphones at the podium and at the teacher desks; it was controlled by a single button that allowed the audio to be muted during small group discussions.

The live-session structure was a 30-minute presentation, followed by 10 minutes of on-site collaboration of teachers in small groups that generated specific questions for the presenter; the presenter responded to questions; the small groups met for 10 minutes to share ideas for incorporating the new knowledge into their classroom practice; finally, ideas were shared with the entire group via the originating site. The participants had one week to complete and submit a post-session, online survey of related CK and pedagogical beliefs.

The video mode used videotapes taken from the live 30-minute presentation broadcast the previous week. The video sessions were aired on a fixed schedule on the same network as the live sessions. The videotape was played by a session host from an origination site to as many as eight remote sites. At the remote sites, a facilitator and teachers viewed the presentation and discussed its content. Next, they were led through a discussion with the other sites, generating questions and considering how they might be able to integrate what they learned into their classroom practice. Finally, the teachers viewed the discussion videotaped during the previous week's live presentation and reformulated the list of questions based on the new information. Specific questions were emailed to the presenter or posted on the project website; answers were posted on the website and emailed to all participants in that video session. The participants had one week to complete and submit a post-session, online survey.

The web mode was distinctly different from the other two modes in several ways. The fundamental notion of asynchronous communication is that it disregards time and place. Participants were flexible in terms of when and where they engaged in these sessions. They viewed a digitized videotape, which used *Macintosh iMovie*®, of the original live session. The teachers interacted with each other through a discussion board within the framework of the Blackboard® course management system. The participants had one week to view the streamed video and interact in the discussion boardroom; they had a second week to complete the post-session, online survey.

The effectiveness of the three delivery modes was examined using a pretest/posttest assessment of the participants' science knowledge and their perceptions of the delivery modes. These instruments contained selected released items from grade 8 National Assessment of Educational Progress (NAEP) and TIMSS and the science content portions of the Praxis National Teacher Examination for elementary teacher certification (http://nces.ed.gov/nationsreportcard/itmrlsx/search.aspx?subject=science), which closely aligned with the presentation topics (Annetta and Shymansky, 2006, 2008). These tests ascertained what participants knew about the science content to be presented in the Y3 ITV schedule and what they learned from the presentations. The pretest results of prior knowledge and dispositions toward constructivist teaching were used to stratify high, middle, and low participants across the three delivery modes.

The pretest, posttests (segregated into post-session tests specific to the topic), and post-ITV session surveys were independently coded and scored by two outside evaluators with content, constructivist theory, and teaching practice backgrounds. They used a four-point Likert scale (0 = no response...3 = thoroughly analyzes and evaluates scientific viewpoints and/or justifies key results or procedures). An inter-rater reliability on constructed response and vignette items was 0.70+. Science knowledge data were segregated into three subscale scores: MC (mean of the five multiple-choice responses), CR (mean of the three constructed-response-item responses), and VIG (average of the two chosen vignette responses). Two-way analyses of variance were used to explore any differences in scores on post-session science content items amongst high, middle, and low prior knowledge of teachers on the three subscores for the three delivery modes. Pair-wise comparisons examined source of difference within each subscale.

The analyses of variance of participants' post-session science scores yielded significant differences ($p < 0.05$) on the MC and CR subscales but not the VIG subscale. Participants in the live mode outperformed participants in the web and video modes on all three subscales, while web mode participants outperformed the video mode participants on MC and CR (Annetta and Shymansky, 2006).

Factors relating to the effectiveness beyond knowledge acquisition were subjected to a maximum likelihood factor analysis with varimax rotation to identify the major scales within the data set based on participant responses (Annetta and Shymansky, 2008). Results yielded the following factors: *My knowledge was enhanced*; *effective interactions occurred*; *technical difficulties hampered my learning*; and *the session built confidence in my science knowledge*. *T*-tests were used to examine differences in the means of the fixed factors for delivery mode and science content. Teachers perceived the live mode to be significantly ($p < 0.001$) more effective than the other modes, and they perceived the web mode to be significantly ($p < 0.001$) more effective than the video mode.

Teacher Interview PD Results

Interviews of ten randomly selected participants were conducted in two parts. First, an electronic survey based on the HRI protocol (http://www.horizon-research.com/LSC/manual/#6) was sent to the target teachers.

From these survey responses, a preliminary analysis identified areas requiring elaboration and more in-depth probing; a customized set of interview questions was developed for each teacher. Second, a telephone or face-to-face interview was scheduled and conducted using the teacher-specific protocol. The electronic survey took 20–40 minutes to complete, and the interview took 15–35 minutes.

ASSERTION #3: **Participants increasingly valued the PD provided; preferred face-to-face delivery; and focused pedagogical or context-oriented content topics compared to pure science topics.**

Teachers were asked to rate their overall experience in the project on a five-point ordinal scale (1 = very negative... 5 = very positive). Results indicated that the teachers interviewed rated their PD experience high to very high over the duration of the project (Y1: *M* = 4.0, range = 3–5; Y3 and Y5: *M* = 4.9, range = 4–5). The advocates within the random sample rated their leadership preparation as high to very high with growing value of the leadership experiences (Y1: *M* = 3.9, range = 3–4 for 10 interviewees; Y3: *M* = 4.5, range = 4–5 for 2 interviewees; Y5: *M* = 5.0, range = 5 for 1 interviewee).

Interview Results for Y1 PD

The K–6 teachers interviewed were all advocates with 60–99 hours of PD. On an average, they rated their experience in the project as 3.9 (neutral to very positive, range = 3–5). The reasons for the overall rating varied; in general, they had positive assessments of the project staff and the activities that applied to classroom practices, instructional strategies, and materials. They believed that the PD was generally effective, but they discounted some of the science content sessions. Technical problems with the ITV networks had negative effects on their rating of the distance component. These advocates were comfortable with their responsibilities and prepared for their leadership roles, and they identified the two leadership workshops as preparing them for these anticipated responsibilities. Several identified the positive role models provided by the summer workshop and project staff. In summary, these advocates rated their collective experiences as worthwhile and as positive influences on their professional growth, classroom practices with the modified learning cycle, and encouragement and support of other teachers to do the same. Their concerns were used by the project staff to adjust delivery modes and clarify expectations regarding the TRBs, mentoring responsibilities, and future plans.

Interview Results for Y3 PD

The K–6 teachers interviewed were a mixture of 2 advocates with 195 hours of PD, 2 second-year participants with 130 hours of PD, and 6 first-year participants with 60–99 hours of PD. On an average, they rated their PD experience as 4.9 (somewhat positive to very positive, range = 4–5). The reasons for the overall rating were slightly varied; in general, they were positive about the PD activities that applied to classroom practices, instructional strategies, and materials but discounted a few science content sessions. The improved assessment of the ITV sessions reflected the adjustments to two parallel schedules for Iowa and Missouri, thereby avoiding the technical interface problems between the ITV networks. These teachers continued to appreciate the opportunities for PD not afforded prior to the project and to collaborate with other teachers in their school district, cooperative, and project who had similar teaching assignments and interests. Their concerns were about availability of the science kits and continued financial support during budget difficulties. The two advocates in this random sample were very or greatly comfortable with their responsibilities and prepared for their leadership roles. They identified the leadership workshop, local field staff, and additional experience as lead teachers as the major factors influencing their self-efficacy. In summary, these teachers rated their collective experiences as very positive influences on their PCK, professional growth, and classroom practices with inquiry science. They believed that they were better prepared to provide quality science instruction but wanted more consideration of assessment issues during the period of increased accountability. Earlier concerns about ITV sessions had been addressed, but some science content sessions were still viewed as being at too high a level for elementary teachers.

Interview Results for Y5 PD

The random sample of K–6 teachers interviewed was distributed across all grade levels, and one teacher was an advocate. The majority had participated in >129 hours (7), 100–129 hours (2), and 60–99 hours (1) of PD. On an average, they rated their PD experience as 4.9 (somewhat positive to very positive, range = 4–5). These teachers valued the experiences with teaching and assessment strategies and inquiry materials but believed that the time to plan and network with other teachers across the school district, cooperative, and project was the most important contribution. The upper grade-level teachers (n = 3) valued the ITV content sessions somewhat more than the K–2 teachers (n = 7). Both groups were very positive about all forms of PD (graduate programs, summer workshops, school-level meeting, peer-led reflective practice sessions, ITV sessions), and they recognized that the availability of these opportunities was a direct result of the project. These beliefs were more apparent in the Y5 interviews than in the Y1 and Y3 interviews, which surfaced potential advantages of the decentralization and cascading leadership features of the project and the focus of the ITV sessions moving away from science content toward PCK and classroom applications of inquiry teaching, assessment, and cross-curricular connections to other disciplines. The single advocate stated, *"I'm more relaxed now that I have learned how to help other teachers."* In summary, these teachers were very positive about the PD and the general results of the project. Clearly, some of the difficulties encountered from external sources helped participants form a stronger alliance and community culture. Furthermore, the cascading leadership and decentralization of responsibilities provided greater local ownership of related goals and activities.

Summary of PD Results

ASSERTION #4: The project was relatively successful in reaching the target PD hours set by NSF and providing high-quality PD focused on CK, PK, and PCK specific to science inquiry kits but did not fully anticipate the turnover in participating school districts' administrative and teaching staffs.

By the project's end, 583 (46 percent) of the 1,269 target teacher population received greater than 129 (set by NSF) hours of PD (compared to 13 percent for all LSCs). The turnover rates for teachers (25 percent), principals (56 percent), and superintendents (67 percent) in rural districts were significant with 1,450 different teachers involved over the 5 years. Further indication of the PD effectiveness is the fact that 102 teachers (Iowa and Missouri) completed M.Ed. degrees with science education specializations at the University of Missouri-St. Louis during the project.

Resources, Topics, Lessons, and Instructional Time

ASSERTION #5: Participating school districts and schools reached the target of providing high-quality inquiry resources at all grade levels, addressing the *less is more* approach of reducing topics covered and increasing depth of coverage; and teachers, schools, and school districts found creative ways to address resource needs.

All districts and schools achieved the benchmark of 14 inquiry-based units in K–6 with very few not having two in each grade level. However, some districts had many more (four–six) science modules at some grade levels; for example, four school districts had three or more modules at each grade level, except kindergarten. Results from the teacher questionnaire indicate that a great number of lessons and time were spent on science over the duration of the project. The survey results of 300 teachers randomly selected by HRI at the start and at the end of the project found that teachers were teaching more lessons per week (3 increased to 3.3) but on fewer topics annually (4.9 decreased to 3.9) for more time per week (114 minutes increased to 120 minutes) during the school year. These results are consistent with the *less is more* theme promoted in science education reforms (AAAS, 1990, 1993; NRC, 1996).

Y1 teacher interviews indicated positive remarks about the instructional materials and strategies introduced in the PD and developed in the TRBs and the support from the school district and the University of Missouri, St. Louis endowment regarding these resources and ideas. Teachers appreciated the opportunities to collaborate and to establish networks with other teachers in their district, cooperative, and project, who had similar teaching assignments and interests. Y3- and Y5-teacher interviews indicated that the PD instructional materials and strategies were aligned with the local science curriculum, and the supportive attitudes of the stakeholders had continued with changes in the school districts and realized budget cuts. Some schools found innovative ways to share, borrow, or rent kits from other schools within their cooperative, Iowa regional educational agencies, or nonprofit groups of retired Missouri science teachers.

Classroom Practice

ASSERTION #6: Classroom practices moved toward the desired approaches and best practices over the duration of the project.

Classroom observations were conducted according to a clinical observation protocol (http://www.horizon-research.com/LSC/manual/#6) that addressed: (a) several demographic, focus, material, and contextual issues; (b) four analytical dimensions (design, implementation, content, and culture) composed of key indicators; and (c) a single holistic consideration. Design focused on planning and organization: goals, tasks, roles, collaborations, and interactions consistent with investigative science; instructional strategies, activities, classroom resources and practices, and time for sense making; and formative and summative assessment and use of this evaluative information in future lessons. Implementation focused on the classroom instruction's alignment with the desired approach and investigative science: teacher's confidence, teaching style, and management and lesson pace and real-time adjustments to reflect students' needs and responses. Content addressed the value, significance, and appropriateness of the science: representation of the nature of science, elements of inquiry as a dynamic process,

Table 3.2 Descriptive Statistics for Classroom Observations

Category	Year One (n = 16)	Year Three (n = 16)	Year Five (n = 16)
Design (1–5)	2.88 (0.81)	3.38 (0.81)	4.50 (0.52)
Implementation (1–5)	2.69 (0.95)	3.19 (0.98)	4.31 (0.70)
Content (1–5)	2.63 (0.81)	3.00 (0.89)	4.38 (0.81)
Culture (1–5)	2.94 (1.18)	3.75 (1.06)	4.63 (0.81)
Capsule Rating (1–8)	3.69 (1.99)	4.69 (1.96)	6.94 (1.24)

connections to real-world applications, and degree of sense making. Culture addressed encouragement and value of active participation, climate of respect, peer collaboration and interaction, academic rigor, constructive criticism, and debate. The mean observation for the analytical dimensions and holistic ratings of 16 teachers randomly selected across the participating school districts for Y1, Y3, and Y5 are summarized in table 3.2. The analytical dimensions of classroom practices (design, implementation, content, pedagogy/instructional materials, and culture) were rated on a five-point ordinal scale (1 = not reflective...5 = extremely reflective of best practices) and overall capsule rating on an eight-point ordinal scale (1 = ineffective instruction/passive learning highly unlikely...8 = exemplary instruction highly likely to enhance students' learning). Mean ratings demonstrate steady improvements in all five analytical dimensions and in the holistic assessment of classroom practice across the project.

One-way analyses of variance revealed highly significant ($p < 0.001$) main effects for all categories. Post hoc pair-wise comparisons revealed nonsignificant ($p > 0.05$) positive changes between Y1 and Y3 and between Y3 and Y5 for all analytical dimensions and capsule ratings but significant ($p < 0.05$) positive changes between Y1 and Y5 for implementation, content, and capsule ratings. Analysis of the observers' field notes indicated that teachers were quick to implement instructional strategies (children's literature as springboards to inquiry, interesting demonstrations, humor, experiments, simulations/role plays, etc.) dealing with the engage and explore phases of the modified learning cycle but were less able to address meaning making, elaboration, and explanation in the consolidation phase (classroom questioning, transformations between representations, trade books, writing-to-learn activities, etc.) and assessment for learning to empower learners and inform instruction in the assess phase. The Y5 observer's comments across the variety of teachers (K, primary, middle years, music, and special education) illustrate these successes and challenges:

The kindergarten lesson was designed specifically to introduce a unit on plants. Using nonfiction literature, the teacher planned to bring out what students know about plants, especially seeds. The investigative nature of the unit was planned to use children's ideas on what it takes to get a seed to germinate. The teacher followed her plan, adjusting occasionally for the individual student who seemed slower with the physical aspects of planting seeds. Overall, the purpose of engaging the students and developing their enthusiasm for the upcoming unit was achieved. The teacher made sure that the scientific terminology was accurate, relating it to terms the students already used at home.

The primary teacher taught a lesson on magnets and magnetism. She used magnets of various shapes and glass containers of paper clips, paper, cloth, and wood so that children could test magnets directly and through the various materials provided. Students discussed their findings, challenged each other's ideas, and finally recorded their findings in a journal.... The teacher's tone was very warm, and students were exceptionally focused and excited. Although not based on any of the LSC-designated materials, the lesson was consistent with the approach of these [materials] and with NSES and Missouri standards for science.

Using three different types of soil, the children in the middle-year class went to stations to examine the soils by feel, smell, and with a magnifying glass. This was the first lesson on this topic following a series on plants. The teacher's intent is to pursue agriculture-related activities and field trips with an environmental emphasis as well as the implications for plant growth and productivity [relating to the local agriculture]. The teacher's design was beyond anything in the kit's teacher guide. It showed creativity and clear knowledge of the investigative nature of pure and applied science. The engagement of students was evidenced by their active and enthusiastic participation and their spontaneous communication with each other.

This music teacher focused on music as a way of bringing out issues of environmental protection. Using slides of earth from space, students sang "Our Spaceship and Our Home" in preparation for their science lessons scheduled for the next day with a different teacher. The teacher interrupted student attempts to ask questions and to offer suggestions regarding environmental concerns. However, the lesson was effective in raising questions, which hopefully will be addressed in subsequent lessons by their regular teacher. Many students were able to make connections and share them before they were squelched. It was clear that the students had knowledge of science as well as enthusiasm for investigation. Unfortunately, the teacher seemed to forget the purpose of the lesson and focused exclusively on skill in performing the song.

This multigrade, special education class used science as the basis for all their study. As part of a social studies resource unit [from a different teacher], students studied "important events in history." The launch of Sputnik resulted in the study of rockets in the science class. Today they will review the Laws of Motion and build a balloon-propelled cart, test it, modify it, predict its speed and distance, and retest several times. The teacher's plan was detailed and complete, using NASA publications and a book entitled "Rockets Plus." The design called for close integration of science concepts with construction and testing of the rocket car. The teacher and the teaching were brilliant in bringing science concepts to the level of students with such varying abilities and energies. Having a carefully considered plan, she departed frequently to engage challenged students. Although students did not interact very much with each other, the teacher was active and successful as a "go between." She had certainly convinced students that she respected their ideas and reinforced that repeatedly with each student.

Student Performance on High-Stakes Tests and Surveys

Assertion #7: The high-quality PD, high-quality resources, and enhanced classroom practices positively influenced district-wide students' science achievement in grades 3 and 6.

The school district, not the classroom or student, was used as the unit of analysis because state high-stakes achievement data were available only at the district level by grade level and student gender. This more conservative approach was not seen as a limitation but as a better representation of the systemic focus of the LSC projects generally and the Science Co-op project specifically (Shymansky, Annetta et al., 2010; Shymansky, Wang et al., 2010a, 2010b). Both Iowa and Missouri report school district scores as a percentage of students passing or advanced. Table 3.3 shows the percentage of students passing and better for the ITBS and MAPS for Y1 (benchmark year) and Y5 (final year) of Science Co-op school districts ($n = 33$; 23 Iowa and 10 Missouri) and selected comparative school districts ($n = 20$; 10 from each state).

School districts selected for comparison were nearby and similar to the Science Co-op districts. These percent scores were converted to standard z scores using sample means and standard deviations of the ITBS and MAPS school district scores respectively (table 3.4).

The resulting z scores for the ITBS and MAPS tests were combined and subjected to a three-way (treatment, gender, grade level) Analysis of Covariance (ANCOVA), using the baseline scores as the covariate to determine the main and interaction effects on the ITBS and MAPS post-project scores. No significant gender effects or two- and three-way interaction effects were found; however, the main effects for PD strategy and student grade level proved to be significant ($p < 0.01$). High-stakes scores across all students in the Science Co-op school districts were significantly higher than those in the comparative school districts, $F(1, 203) = 20.80$, $p < 0.001$, partial $\eta^2 = 0.093$ (a medium effect size, according

Table 3.3 Scores on the ITBS and MAPS by Grade Level, Gender, and Year for Participating and Comparison School Districts

	Grade 3				Grade 6			
Science Co-op Districts	**Male**		**Female**		**Male**		**Female**	
	M	***SD***	***M***	***SD***	***M***	***SD***	***M***	***SD***
Year One								
ITBS	67.57	11.50	68.39	9.90	66.91	10.28	64.91	11.20
MAPS	27.90	13.34	25.00	13.69	13.80	11.04	4.80	4.54
Year Five								
ITBS	79.30	8.86	81.96	6.31	80.00	6.44	76.09	7.49
MAPS	58.50	10.91	51.30	14.32	24.80	19.70	19.30	13.06
Comparison Districts								
Year One								
ITBS	76.40	5.54	74.50	18.27	76.80	12.10	70.10	10.35
MAPS	67.30	14.68	52.70	22.89	29.50	14.51	24.10	9.98
Year Five								
ITBS	73.30	8.45	72.70	9.45	71.50	8.95	70.50	5.06
MAPS	55.10	11.63	50.70	14.17	38.90	18.92	16.90	9.50

to Tabachnick and Fidell, 2007). In addition, high-stakes scores for grade 3 students in both school district groups were significantly higher than those for grade 6 students, $F(1, 203) = 19.07$, $p < 0.001$, partial $\eta^2 = 0.086$ (also a medium-effect size). Male and female students in both grade levels performed similarly. The positive high-stakes test results, which serve as the basis for state and local decision making, suggest that an adaptation strategy and a combination of regional live workshops and ITV with ongoing local leadership, advocacy, and support like that utilized in Science Co-op can serve as a viable PD option for K–6 science.

Discussion

The results of the classroom observations, PD assessments, and teacher interviews were not surprising because the project emphasized the successful features of Science PALs (TRBs that adapt inquiry modules with reform standards, children's ideas, cross-curricular connections, modified learning cycle focused on science literacy) applied to these isolated Iowa and Missouri rural school districts and all K–6 teachers, not just

Table 3.4 Standardized Means (z) and Standard Deviation for Scores on ITBS and MAPS by Grade Level, Gender, and Year for Participating and Comparison School Distircts

School Districts	Gender				Grade level			
	Male		Female		Grade 3		Grade 6	
	M	*SD*	*M*	*SD*	*M*	*SD*	*M*	*SD*
Year One								
Science Co-op	−0.25	0.85	−0.36	0.86	−0.13	0.82	−0.48	0.85
Comparison	0.70	0.91	0.30	1.11	0.90	1.06	0.10	0.83
Year Five								
Science Co-op	0.24	0.97	0.10	0.94	0.52	0.85	−0.17	0.94
Comparison	−0.10	1.03	−0.46	0.95	0.07	1.02	−0.64	0.86

science teachers (Shymansky, Yore, and Hand, 2000; Shymansky, Yore, and Anderson, 2004; Yore, Anderson, Shymansky, 2005). Most of the participating teachers were generalists, and very few had strong science backgrounds or specialties at the outset of the Science Co-op project. However, science literacy—in fact all disciplinary literacies—are a whole school responsibility, not just the specific content specialists (Yore et al., 2007). The problem-centered, adapt-a-science-kit was viewed by participating teachers as authentic professional learning and necessary teacher work, which built on their strengths and teaching experiences rather than emphasizing their science content weaknesses—often the case in many PD programs and the emphasis of the LSC-funding program. The teacher networks encouraged by project-wide meetings, multiple district regional workshops, and ITV sessions were viewed positively and as an effective strategy to overcome a teacher's classroom-bound isolation caused by the rural geography and student responsibilities. Collaboration with other teachers with the same teaching assignments allowed them to explore the science content needed to teach the kit *in context* and the engage-explore-consolidate-assess teaching model provided useful structure on which to plan and implement more achievable inquiry teaching, which reflected the subtle, grade-specific differences in priorities and development frequently overlooked by a global elementary or middle school perspective. These efforts went well beyond mechanistic familiarity with the equipment and activities to pedagogical-content foundations of the inquiry units in context of real classrooms and target learners (Yore et al., 2007). Teachers quickly enacted children's literature, demonstrations, and games to access and challenge prior knowledge and engage students and a variety of activities to explore these ideas; however, they were somewhat slower to realize the importance of postexploration classroom questioning, negotiation, and deliberation to make sense of these experiences and construct conceptual understanding. Assessment for learning—as contrasted to assessment *of* learning—was accepted as an important contribution to empowering learners and informing future instruction. The teachers were also receptive to other meaning-making strategies (e.g., concept mapping, science notebook, and writing-to-learn tasks) and other disciplinary literacy activities.

Despite the challenges, participating schools established communities of practice that encouraged groups of teachers to work together for mutual gains and facilitated desired classroom practice with adapted inquiry science kits. These teachers were empowered by the cross-curricular connections and were comfortable with the modified learning cycle model for inquiry science teaching promoting science literacy. They were able to embed language activities in the inquiry modules to more fully reflect the fundamental sense of science literacy and the functionality of language and literacy tasks in constructing understanding, persuading others, and reporting knowledge claims while the semi-structured teaching approach provided some boundaries for classroom management and inquiry. The classroom observations indicated a rapid adoption of the engage and explore strategies and steady improvement in the consolidation strategies and the meaning-making process. Assessment was noticeably improved, but additional consideration would be useful.

The positive, but tentative, effects of the high-quality PD, locally adapted inquiry resources, and classroom practices on students' science achievement were pleasing and helped support the relationships within the LSC theory for action (Czerniak et al., 2005; Geier et al., 2008; Revak and Kuerbis, 2008; Johnson, Fargo, Kahle, 2010). More importantly, the supportive evidence for this claim was in the currency of local decision makers and teachers. ITBS and MAPS scores are the information used in the accountability process of most Iowa and Missouri school districts and part of their established annual-review framework with the state departments of education as part of *No Child Left Behind*.

The science advocates' responses were very positive and indicated growing comfort as the project unfolded. The advocates and partners formally and informally reported to the lead evaluator that they had developed a level of confidence and recognized expertise as an inquiry science teacher and to be an effective mentor and PD provider. Self-efficacy is difficult to develop as teachers are slow to accept their own expertise as models and facilitators of professional practice.

Closing Remarks

The legacy of passionate, well-educated advocates and ongoing leadership for science education is highly valued and much needed in rural America. These lead teachers were the lasting legacy of Science Co-op in the rural school districts as advocates for high-quality science education programs, which parallels the well-established advocacy for athletic and music programs. The value of the hybrid delivery system for PD, ITV applications, community-university partnerships, and cascading leadership were implemented using existing technologies and proven models; as well, its practical applications have illustrated how high-quality PD can be provided and sustained across isolated school districts with limited financial and expert resources. The *co-op* solution to resources in financially challenged districts, where teachers set up sharing and delivery systems for neighboring districts and rental systems involving a state retired teachers association and area education agencies are examples of rural ingenuity. Furthermore, the same determination was found in how regional clusters of districts networked and shared teachers and local resource people from rural industries and government agencies to enhance many PD activities. We celebrate these schools' and teachers' successes and believe that they can be replicated in other rural systems and subsystems.

Technological advancements have increased the power of ITV, which has begun to morph into more sophisticated and elegant platforms, such as 3-D virtual learning environments (Annetta, Folta, and Klesath, 2010). Through these environments, PD participants can be exposed to real-time interaction while gaining the online presence once only acquired through ITV. Furthermore, recording the PD events and user interactions becomes easier as self-regulated learning can be virtually observed through server-side data collection. Finally, 3-D environments allow those who lead the session to create simulated scenarios where participants can practice what they learn without the need to send and manipulate materials to distant locations. Scenarios such as these are closely aligned with traditional classroom or workshop instructional practices because interactions and activities are seamlessly integrated into the instruction, and the practice can be replicated once it is constructed. As these ICT continue to become more user-friendly, user-created content (e.g., constructivist learning) becomes more of a reality. Where once it was left to the PD developer to create the activities by which participants interact, now participants create their own activities within these environments, allowing for real-time assessment of learning to empower participants and inform adjustments to instruction.

The financial, personnel, and time demands of scaling proven, short-term (one–two years), single-site PD programs to larger, long-term (five years), multischool or multidistrict systems were not well understood by funding agencies and project staff at the outset of the Science Co-op project. The LSC projects did recognize that planned change is a process—not an event—that takes time to establish trust and local sponsorship, to provide PD, to build leadership capacity and ownership, and to ensure teacher uptake. However, most people involved in funding and planning treated the scale-up procedure as a linear function when in fact it was an exponential equation. Funding was allocated as simply the number of teachers multiplied by a set dollar value per participant predicted to reach 130 PD hours. Unfortunately, this formula did not consider geographic coverage, multiple authorities, and other dynamics within and across sites located in two different states. Changes in school district administrative staff and participating teachers made it difficult to maintain momentum, establish steady state in participation, and reach the 130-hour PD goal for each teacher. The most surprising difficulties encountered were changing NSF program staff, LSC intentions, and interpretations by the NSF. Changes in program directors and officers were associated with changes in intentions, expectations, and supports for a peer-reviewed, approved, and funded project. Furthermore, the stress placed on teachers to collect data, some of which had little relevance to local needs, by the LSC-wide evaluator and the additional demands of collecting student-achievement data by the project evaluator consumed social capital at a rate greater than it could be developed by project and regional staffs.

Chapter 4

Lessons from the Field: Examining the Challenges and Successes of a Mathematics and Science Program Using Acceleration and Enrichment for Gifted Urban Middle School Students

Toni A. Sondergeld, Andrea R. Milner, Laurence J. Coleman, and Thomas Southern

It has been clearly established over the last quarter of a century or so that science, mathematics, and technology education among precollege students in the United States is severely lacking (National Science Board, 2006; Carnegie-IAS Commission, 2009). Problem solving, analytical skills, and critical thinking associated with these content areas are necessary traits for our students to possess in order for the United States to successfully compete in the twenty-first century. While our nation at large struggles to provide quality K-12 education in mathematics, science, and technology, this situation is even more bleak for urban and low socioeconomic status students (National Science Board, 2006; National Center for Educational Statistics, 2009). At-risk children often face additional barriers including lower-quality teachers who are not adequately trained to teach mathematics and science (National Comprehensive Center for Teacher Quality, 2006); limited advanced- or rigorous-course offerings in these content areas (Norman et al., 2001); and insufficient materials to support the needs of quality instruction (UNESCO, 2008).

So what is being done to bring quality science, mathematics, and technology education to all K-12 students in the United States? At the national level, mathematics and science standards have been created or are in the process of being revised to help level the playing field by providing rigorous expectations for all students. Teachers are being held more accountable for student learning in these content areas through standardized tests in mathematics and more recently science. Various agencies, federal and private, are supporting grants focusing on science, mathematics, and technology education. One such program, Project Excite served as a catalyst for our project (Olszewski-Kubilius, 2003).

The Accelerating Achievement in Mathematics and Science in Urban Schools (AAMSUS) award was given to the University of Toledo in 2004, a federal grant supported through the Javits Gifted and Talented Students Education Program. The purpose of AAMSUS was to raise student achievement and improve attitudes toward mathematics and science of economically disadvantaged, limited English proficient, and/or disabled learners who have potential for more advanced achievement in mathematics and science. A primary goal of AAMSUS was to identify at-risk children with high potential early and provide accelerated instruction in the curricular areas of mathematics and science. Children were identified under three conditions (gifted/scores, scores combined with teacher recommendation, and recommendation only). These groupings were to determine the impact of the intervention with the various groups. To impact urban students most, early identification is a key (Coleman and Southern, 2006). The out-of-school environment is often unsupportive of academic achievement among minority populations (Ford, Grantham, and Whiting, 2008), and being academically successful may be seen as a betrayal of one's roots (Rowley and Moore, 2002). These destructive attitudes begin in elementary school and make it almost impossible to address by the time students are in middle or high school. Therefore, if AAMSUS was to have a chance of helping urban students develop an identity of "mathematics achiever" or "science learner," identification for the program needed to be completed early. Thus, children in the AAMSUS program were identified in their fourth-grade year to participate in the program from fourth through eighth grade. Additionally, to combat negative attitudes about learning and motivate these at-risk young students to engage in learning through AAMSUS, creating a support network (trained professionals, a cohort of students, and mentors) so that student academic talent could flourish was another essential priority of the AAMSUS program.

The primary intervention was to provide students with accelerated experiences (and enriched experiences as needed) within a rich curriculum.

Enriched experiences are horizontal activities in which children learn elaborations of content at the same curricular level. Accelerative experiences are vertical activities in which children learn more advanced content beyond their grade level. The accelerative principle underlying the intervention was to maximize time spent on new challenging experiences and minimize time spent on known information and skills. Acceleration has been shown to be a highly effective strategy for meeting the needs of gifted students (Kulik and Kulik, 1984; Rimm, 1986; Rogers, 1991). The use of acceleration for students in summer and extracurricular programs, like Saturday seminars, is also well documented (Benbow and Stanley, 1984). Accelerative options are efficacious for academic achievements (Southern, Jones, and Stanley, 1993) because acceleration has the advantage of being tied to state content standards that are integrated into the curriculum directly. Acceleration allows students to spend increased time learning new content without merely engaging in drill and recitation, and allows the integration of academically relevant enriching experiences that are linked directly to content mastery. Until now, acceleration programs have been availed by students from advantaged backgrounds who already have had generally enriched and accelerated backgrounds that prepare them for advanced experiences (Council of the Great City Schools, 2003). Project AAMSUS focused specifically on providing accelerated content to less-advantaged children in order to prepare them for participating in advanced academic mathematics and science courses in high school and postsecondary education. Further, the relationship between acceleration and the development of talent has clearly demonstrated that participation in a domain is an essential component of advanced development (Bloom and Sosniak, 1985; Feldman, 1994; Ericsson and Lehmann, 1996), and underprivileged urban students simply do not have sufficient opportunity to engage in such programs (Council of the Great City Schools, 2003). Thus, the AAMSUS program, if implemented effectively, should have helped to fill a gap in educational inequality for the educationally at-risk students involved.

AAMSUS Program Specifications

AAMSUS was designed to change the academic opportunities available by reducing the time spent on content already known and inserting learning opportunities to increase achievement in science and mathematics that are typical for educationally advantaged children. To do this, students identified as high potential by their teachers in fourth grade participated in the AAMSUS accelerated mathematics and science program through their eighth grade year, providing them with five years of advanced course work in these content areas. Implementation of academic instruction took place in two phases each year. Phase 1 consisted of ten five-hour Saturday sessions that took place every two to three weeks during the course of the normal school year. Students attended mathematics, science, technology, and additional enrichment classes at the University of Toledo. Phase 2 was a two-week summer residential program where students engaged in the same content but remained on campus from Sunday evening through Friday evening for two consecutive weeks. Each year, Phase 2 program learning ended with a culminating educational field trip. The main purposes of these field trips were to reinforce learning through authentic learning opportunities and motivate students to return to the program in the following year. Table 4.1 depicts the specific content and educational field trips of AAMSUS in which students participated. While the main goal of providing academic accelerative opportunities for at-risk, high-potential students in mathematics and science drove program decisions and implementation, the story we share clearly depicts the practical and somewhat unexpected challenges related to implementing the AAMSUS program with urban preadolescents as well as the successes despite the yearly obstacles.

Challenges to Implementation

Implementation problems or issues that developed in the first year of the AAMSUS program in the Toledo region foreshadowed the limitations of the program implementation for the subsequent four years. These fell into three main categories: (1) logistical issues connected primarily to the peculiarities of the Toledo region in the time period of the project; (2) problems with teacher apprehension for teaching more advanced, out-of-grade level content through unfamiliar methods; and (3) population-programmatic issues connected to the interaction between an accelerative-oriented program and the students and/or parents involved in the program.

Logistical Challenges Impeding Implementation

The public school system that the University of Toledo worked with is an old urban system trying to innovate within a web of social, economic, and political issues that impedes change. We will refer to this system as the Urban Public School System (UPSS). Logistical challenges were influenced by four factors: date of implementation, dire economic circumstance and diminishing resources, rigid rules, and bureaucratic mentality. Date of implementation refers to the time between grant application and granting the award. During that time frame, key personnel in the school system changed. Thus, understandings among the parties were muddied and incomplete and consequently had to be renegotiated and reestablished.

Recruiting Teachers as Cohort Mentors

The project was based on finding interested fourth-grade teachers in schools and using their classes and schools for the succeeding years. Participating schools were to be in one or more catchment areas served by a particular middle school because we would be working with the same students for five years, and in the last year students would be combined into the middle school serving those elementary schools. Original teachers and their colleagues who joined in subsequent years were to form a cohort of support for project students and new teachers joining the project. Thus, teacher recruitment preceded student recruitment.

Recruiting teachers required getting permission from the principal and persuading teachers to join our project. At the time we began teacher recruitment, the UPSS teachers' union and the principals' (or administration) were at odds. The teachers' union had veto power over teacher involvement; the principals controlled entry to a school and had limited recourse over teacher decisions; yet both had to agree. The teachers' union was at that time "working to the rule." Teachers expressed a sincere belief that the "kids come first," and they would protect them at all costs. Administrators echoed this sentiment at the central office and in the local schools. With the lack of trust between the groups, the relationship between teachers and administrators had hardened into rigid rules. The "kids come first" mission seemed to get pushed down the priority list. Everything took an improbable amount of time, and "no decision until we talk to our group" was the prevailing mantra. Months went by before we got actual permission to go into a school and speak with teachers for recruitment. The intention was to recruit and add teachers each summer to the cohort on a yearly basis.

Across the discussions among the organization representing the groups, the notion of how the project would benefit children and teachers was assumed and rarely mentioned; rather it was about what was appropriate compensation for atypical work per union contract. In our trips to individual schools to invite teachers to join, we discovered that we had to compete with other projects from rival universities who had more compensation to offer than we did for teacher time. The impression communicated by this

Table 4.1 AAMSUS Content and Educational Field Trips by Year

Year	Student Grade Level	Phase 1: School Year Content	Phase 2: Summer Content	Culminating Educational Field Trip
One (2004—05)	Fourth	**Science:** Great Black Swamp Theme—air, water, and Earth lab experiments. **Tech:** Microsoft Word and Internet searches as tools in scientific inquiry. **Mathematics:** Frequency, forming categories, and probability related to study of Great Black Swamp.	**Science, Tech, and Mathematics:** Continuation from Phase 1: Great Black Swamp Theme	**Monsoon Lagoon Water Park**—students conducted small experiments with water and enjoyed the water-park attractions.
Two (2005—06)	Fifth	**Science:** Ten scientists visited (e.g., chemist, structural engineer, forensic psychologist), spoke about their career, and then led students in related lab. **Tech:** Internet skills used to study more about lives of scientists in fields they were interested in. Microsoft PowerPoint learned. **Mathematics:** Geometry, measurement, and statistics	**Science:** Biology was the main focus with time split between zoology and botany during the day. Evening science split students between simple machines and astronomy. **Tech:** Internet skills used to study more about scientists' in students' field of interest. **Mathematics:** Geometry, measurement, and statistics	**Put-In-Bay Island in Lake Erie**—students learned about catamarans, caves, geodes, minerals, and butterflies.

Three (2006—07)	Sixth	**Science:** Problem-Based Learning unit on Electricity—Students worked to solve the problem of a power outage in the community. **Tech:** Internet research linked to PBL unit. **Mathematics:** Problem-solving skills linked back to the PBL unit.	**Science:** Complete electricity projects where students met an electrical engineer, electrician, and built models. Additional courses with lab experiences were also available in physics, chemistry, biology, and psychology. **Tech:** Internet research related to science content including local nuclear power plant. **Mathematics:** Algebraic skills/ problem solving.	**Davis Bessie Nuclear Power Plant**—watched electricity being produced. **Local Wildlife Preserve**—observe bald eagles.
Four (2007—08)	Seventh	**Science:** Students grouped by science-interest groups to plan and carry out science experiments focusing on their special-interest topic (e.g., weather, music, animals, sports, forensic science).	**Science:** Roller Coaster Science—physical science concepts such as kinetic and potential energy, friction, inertia, force, et cetera.	**Cedar Point Amusement Park**—conducted experiments created by NASA that were designed to be done at this specific theme park.

Continued

Table 4.1 Continued

Year	Student Grade Level	Phase 1: School Year Content	Phase 2: Summer Content	Culminating Educational Field Trip
		Tech: Internet skills linked to researching science topic of interest with small group. **Mathematics**: Algebraic skills/ problem solving.	**Tech**: Microsoft Office and Google Earth. **Mathematics**: Algebraic skills and problem solving.	
Five (2008—09)	Eighth	**Science**: Individual science fair projects researched, conducted, and presented based on topics of interest. **Tech**: Related to researching for independent science fair projects and creating reports. **Mathematics**: Algebraic skills/ problem solving.	**Science and Tech**: Continuation from Phase 1—Individual science fair project completion and presentation to expert panel of judges. Mathematics: Algebraic skills/ problem solving.	**Washington, D.C.**—visited the Smithsonian Institution, the Library of Congress, and the Capitol to meet our state representative.

process was that teachers and principals were looking for the "best deal" among competing opportunities.

Sixteen schools were initially visited; eight schools had teachers who wished to participate in the first year. The plan was to continue in those eight schools and recruit new teachers each succeeding year to join their colleagues. This would continue into middle school with a new set of teachers. Ultimately we ended up with 12 teachers to begin with in the first year. Each teacher was a general classroom teacher with an elementary teaching license. Their experience ranged from 3 years to 25 years. Teachers were primarily female with only 3 of the 12 teachers being male, which was considered typical for an elementary school. Regardless of teacher demographics, all teachers involved wanted to participate in AAMSUS and committed to teach in the program for one year and to be available to coach colleagues in subsequent years. However, as will be shown in the teacher challenges section that follows, there were difficulties with not having enough structure for cooperating teachers, lack of content knowledge, and motivation for participation.

Identifying and Selecting Students

Once the teachers were selected, they participated in an identification and selection process using the students in their classroom. As part of the process, the teachers were trained how to identify children as gifted based on their behaviors in the classroom. Student selection was completed a week later using student test scores and teacher records. Children were then selected for the AAMSUS program on the basis of three criteria: identified as gifted or attaining high scores on standardized tests alone, test scores (lower) combined with teacher recommendation, and teacher recommendation alone. Three groups were formed of an approximately even size. The numbers in each category were 33 (gifted/scores), 37 (scores and teacher recommendation), and 31 (teacher recommendation alone). It should be noted that the specific classroom and teacher were the context for the selection so that we could see how the intervention would work in a real-world general education class. The unit of study was the students in a teacher's class, rather than the school. Thus, students in two of the groups (scores combined with recommendation or teacher recommendation alone) were unlikely to have been identified by an expert in the field. In one class, the teacher nominated the majority of her students for participation in AAMSUS as she held the belief that most of her students were likely to be gifted. Being that a goal of the program was to allow teachers to identify those students they believed had high potential, we accepted all students nominated by their teachers. While we believe that all children deserve challenging, enriched educational experiences, the selection of some students with less academic ability proved to be problematic for program implementation with the academically advanced nature of the program's content.

Parents and guardians were invited to "get acquainted meetings" at alternate times and dates with the project administrators and the parent liaison coordinator. No parent refused permission to join the program. Slightly more than 60 percent of the families attended while the remaining parents gave written consent without attending a meeting. Parents expressed their hopes for their child's future, their safety concerns about being on a university campus, et cetera. The pattern of parent attendance at meetings decreased through the five years to average 40 percent with some families attending 100 percent of the meetings over the duration of the program. Parental involvement and support for their child participating in AAMSUS will be discussed in greater detail in the section focusing on student and parent implementation challenges.

Student Transportation Issues

Anticipating the difficulty of having urban elementary students attending classes on Saturdays because of distance to the campus, access to transportation, and familial logistical issues, the project provided transportation to the University of Toledo campus. Two scheduled bus routes were used to pick students up at neighborhood schools and return them at the end of the program. Each route took about 40 minutes. The bus went in one direction and reversed itself in the afternoon so waiting was balanced. On the surface this seemed a sensible way to retain student participation in AAMSUS. However, it turned out to be problematic. To keep transportation costs low, we utilized university buses with licensed university-student drivers. Hidden from view was a policy that rotated student bus drivers so a new driver was used each Saturday. This resulted in confusion and delays. Drivers were often late, misunderstood the bus route, and did not wait for students if the bus arrived early. All of this disorganization was amplified by the inability of university-student drivers to adequately supervise children. As a result, we added teachers on the bus to insure more consistency. Even so, this meant that we were never able to standardize the transportation schedule. From the first Saturday to the fiftieth Saturday, five years later, the Saturday busing situation was in constant flux.

Teacher Challenges Impeding Implementation

As previously stated, the AAMSUS program's vision was to recruit interested fourth-grade teachers in schools and include their classes and schools for succeeding years of the program. Teachers and their colleagues who joined in subsequent years were to form a cohort of support for the project, students, and new teachers. The purpose of forming this cohort was to provide academic consistency throughout the five-year program for the students as high teacher turnover rates result in a deficit of quality teachers and instruction; a loss of continuity and commitment; and time, attention, and funds become devoted to recruitment versus support (Brown and Wynn, 2007).

Teachers Lack of Pedagogical Content Knowledge

Whole-group instruction is a model of instruction that is widely used in urban schools (Thompson, Ransdell, and Rousseau, 2005). This instructional method is teacher centered and didactic as the teacher presents lessons to the whole class and all students complete the same assignments at the same time (Stepanek, 1999). AAMSUS wanted to target differentiated instructional strategies that focused on challenging and accommodating the needs of urban gifted middle-grades students as differentiating instruction incorporates best practices while moving all students toward proficiency (Anderson, 2007). As a result, teacher training on differentiated instruction was conducted prior to the first year. Every teacher used this as an opportunity to meet their contract requirements for professional development. Teachers took six hours of course work at the University of Toledo:

- ***Course 1*: *Introduction to Talented and Gifted Education*** (three credit hours)
  A survey of major topics about the education and development of talents and gifts, including identification, social-emotional development, curriculum, creativity, programing and evaluation while interpreting the topics in the teachers' context of the UPSS; and
- ***Course 2*: *Practicum in Talented and Gifted Education*** (three credit hours)
  Field experience to use and refine the strategies for persons with talented and gifted abilities in which they developed and taught accelerated and enriched curriculum.

The University of Toledo provided 1/3 reduction in tuition for project teachers. During this training, teachers were taught how to shift from whole-group instruction to differentiated instruction with flexible groupings, group children according to instructional needs, and use accelerative/enrichment strategies in their home classrooms as well as on campus.

For example, teachers were taught to utilize curriculum-based pre-assessments to find out what the students knew or did not know, and then adapt instruction based on students' immediate academic needs. Teachers also developed inquiry-based lessons within the themes of the project curriculum and used their new skills in the program. One aspect of the grant was the training of teachers to deliver instruction in the Saturday and summer program and to carry that over into their classrooms.

Regardless of the pedagogical training, AAMSUS teachers were worried about teaching more advanced, out-of-grade level content from the beginning. According to yearly debriefing interviews with the teachers, there was low teacher self-efficacy and concern as to the impact this would have on AAMSUS students. Teacher self-efficacy is generally defined as a teacher's belief in his or her skills and abilities to positively impact student achievement, while general (outcome) teaching efficacy has been defined as a teacher's belief that the educational system can work for all students, regardless of outside influences such as socioeconomic status and parental influence (Swackhamer et al., 2009). Students of teachers with high levels of self-efficacy have been shown to outperform students of teachers with lower self-efficacy including rural and urban students (Swackhamer et al., 2009). With all of this in mind, in the first year, there was extensive teacher coaching from curriculum consultants. Also, teachers were placed in teams so that they could help each other on campus and would be able to interact with colleagues in the home school in a similar manner. This concept of team teaching was the intended mode of instruction in order to change the situation from an individual to a group learning and problem-solving process.

Again, based on information gained from yearly debriefing interviews with teachers, team teaching did little to improve their self-efficacy. The teachers were pleased with how much more they thought their students learned, how misbehavior decreased, and one said, "I felt like I was really teaching!" Despite this good news, teachers stated in follow-up interviews that they would not use these materials or methods in their home school classes the next year because it was too much work and did not fit in their schools curriculum. Debriefing at the end of the first three years was the same: teacher self-reports of advanced mathematics and science instruction and computer skills beyond their content ability and comfort level.

Additional School District Complications

Teacher pedagogical challenges were amplified by the circumstances of the school system. The area was in crisis, and resources were diminishing. To deal with these exigencies UPSS began to close schools and reassign teachers and students in the second and third years of the program. These unforeseen changes of AAMSUS disrupted implementation and altered the research design as we lost many teachers each year because they were dispersed to multiple schools. Teachers in original participating schools who earlier agreed to participate in AAMSUS were displaced due to seniority, lost their jobs, or simply declined to participate any longer. By the end of the third year we were serving children in 17 schools instead of the original 8 schools, and none of these schools had AAMSUS trained teachers in them.

Ultimately, the idea of a cohort of supportive teachers was abandoned. While this was highly unsettling, it enabled us to directly confront the persistent and growing problem of the lack of teacher content knowledge. By the fourth year of the program, the teacher role was redefined with the approval from the federal funding agency program officer. Faced with no AAMSUS teachers in the children's schools, two new teacher roles were created: Liaison Teachers and Content Specialist Teachers to continue over the last two years of the project. Three previously trained teachers functioned as liaisons to maintain continuous contact with children and families regardless of school location. Liaison teachers contacted parents the week before Saturday program meetings to remind them when and where to be for bus pickup, helped manage children's behavior, and taught enrichment courses that they felt comfortable teaching in the program. Content specialist teachers were hired for science, mathematics, writing, and technology.

Student and Familial Challenges Impeding Implementation

Students and their families also posed great challenges to implementing AAMSUS. Parental communication, student retention, and student behavior are examples of the barriers we faced during the program. While this next section will elaborate on these sometimes complex struggles, it is vital to remember just how important this program was to some of these children. To exemplify the importance of AAMSUS a brief anecdote about one of our former AAMSUS students, Jamal, is shared here before going into the vast challenges we faced.

> In the spring of 2005 we met students on Saturdays on the university campus. One young boy, Jamal, in the first week was so excited by the program he spoke about it all week in his homeroom. The next Saturday he was absent. His teacher was surprised and asked what happened. Jamal told her apologetically that he overslept. His teacher arranged for him to have an alarm clock. The third Saturday he appeared looking exhausted. When

> asked, Jamal said his parent would not let him set the alarm because it might wake others. So, he stayed up all night in order to not miss the bus. Jamal needed AAMSUS. He attended for 3 years and voluntarily spoke about his interpersonal difficulties with his peers because of his academic interests. For two years Jamal never missed a class. His family left the city [after three years of program involvement] for another state to seek employment (L. J. Coleman, personal communication, May 27, 2010).

This anecdote illustrates how AAMSUS identified real educational needs, engaged students, and how external factors often influenced the outcomes.

Familial Programmatic Issues

Many familial issues challenged the implementation of AAMSUS and student learning outcomes. Problems included communication (getting and maintaining accurate phone and addresses; face-to-face meetings with parents); child-rearing issues (inaccurate perceptions of university setting, protectiveness of children, and child's place in the decision-making process in the family group; retention); and student behavior during the program, which will each be addressed more fully in the sections that follow.

Communication Issues. Communication between the families and AAMSUS was always in flux and never settled, thus making continuous communication between families and program officers a persistent problem. Mail and phone contacts information was often incorrect. Mail would be returned, or we would be told by the family that the mail had not reached them. Phone numbers would often be out of service. To combat these issues, we devised various strategies to keep the lists up-to-date and accurate, though none were satisfactory. Around the fourth year, the AAMSUS person responsible for keeping records noted that earlier addresses were reappearing. We were never able to get families to inform us of changes; even students were often unsure. While problematic, these issues were not entirely surprising given the transient nature of low-income urban students (Stahl Ladbury, Hall, and Benz, 2010).

Face-to-face meetings with families were problematic. Announced meetings, held on multiple nights and in different venues to help people attend, generally had low attendance. Our highest attendance occurred at the first recruitment meeting of the project. As a result, we decided to hold meetings in the summer when families dropped their children off for the residential programs. At those times we informed family members of their child's progress, future AAMSUS plans, and asked for feedback. Face-to-face meetings with families in the form of focus groups for more specific feedback were problematic as well. Though family members confirmed and project staff reminded them of the meeting, most families did not attend. An additional meeting time was added during summer drop-off time when we knew we would have a better chance of interacting with parents.

*Student Rearing and Retention Issues.*The specifics of this theme weave together in complex ways. Families wanted the best for their children, and thus the opportunity to attend AAMSUS was welcomed. All parental figures expressed gratitude for the invitation and expressed hope of their child entering postsecondary education.

Overall, few parents had pursued postsecondary study, so their portrayal of campus life was largely from images of popular media leading some to believe campus was a place where students engaged in behaviors such as drug use, sexual adventures, and dorm invasions. They viewed these fearful misconceptions as potential dangers for their adolescent, and parents wanted to be certain their children were protected. Like all fears, they had to be addressed repeatedly. A good example of this thinking was a family well known for antisocial behavior in their community who were adamant that their child be protected from college students while attending the AAMSUS summer program.

The child's role in the family decision-making process influenced the program and student retention. We saw repeated instances when the adult's desire to have the child in the program was met by a child's wish to leave because of some minor issue, such as "my roommate does not like me," "the food is bad," "I am tired," "I have trouble sleeping," or "I miss my cell phone." These moments were familial decision points. We witnessed many parents conceding to the child's wishes even though the parent said to us that they did not want that outcome. In relatively few instances the family, usually with the grandmother's support, denied the child's request. The minor uncomfortable incident passed, and the child stayed in the program. Another variant of this situation was parents stating that they missed their absent son or daughter. "My baby has never been separated from me for so long." Parents asked if their child could sleep at home and attend during the day in the summer residency program portion of AAMSUS. We did not honor that request because treatment fidelity meant that children participated in dorm life as well as academic life. For some families this was initially acceptable, but if the child showed the slightest sign of discomfort, AAMSUS was abandoned. It appeared that the parents allowed their children to decide whether they stayed in the program. In the first year, most children who had the greatest difficulty adjusting eventually left the program. In subsequent years, decisions were less visible. Children simply did not return to the program. Families would not respond to requests for exit interviews, and the other students provided us with their notion of the child's leaving. Interestingly, four children asked to return after leaving for a year. They all returned and left again in a later year to not return. Student retention was a problem we never solved.

Student Behavioral Issues. Behavior was an issue that appeared to have roots in students' prior experiences in school and home. These at-risk students rarely, if ever, had opportunities in their home schools to be independent learners. They were taught in whole groups with direct instruction and had learned a rote orientation to learning as common with many urban learning environments (Manning, Lucking, and MacDonald, 1995). AAMSUS students expected learning to come quickly and regarded it as a sign of being smart. Asking students to make choices within a challenging curriculum and set their own standards for the quality of work was taxing for them. It took the entire five years to move them from whole group to small group and finally to independent work. This was in part due to student resistance to instructional change and forced thinking, and in part because of teachers' lack of content confidence until later years of the program when content specialist teachers were introduced. Once independent work began, it was not uncommon to hear complaints about working alone. Some children still had great difficulty on the final independent science project, since the learning task was so different than their traditional educational experiences. Furthermore, our students reported having much free time and minimal supervision by adults of their out-of-school environments. AAMSUS provided for very little free time while at the program. The plan for dealing with inappropriate behavior was to have students involved in engaging activities throughout the days and expect them to be responsible for their actions and learning. Not all students welcomed this new kind of freedom. Although the students had chosen to attend the program indicating interest in academic learning, at the same time they were unsure of the seriousness of the enterprise and had mixed feelings about their personal goals. Some students used AAMSUS as a place to fulfill their social rather than academic goals. Thus, students were in an atypical academic environment that continually encouraged them to be independent learners, and their inappropriate behaviors were a consequence of their difficulty in meeting that demand and their adolescent needs.

Successes Despite Challenges

As with all federally funded educational grants, external evaluation of the program's impact on students must be assessed. Most commonly, grant evaluators are required to assess the academic outcomes of students because this is typically of greatest importance to the federal government and provides justification for continued funding. For this reason, AAMSUS's top evaluation goal was to assess whether or not student achievement in mathematics and science had increased over the course of the program. While we believe this is a very important measure of a program's success, it is also essential to measure a program's impact on related affective domains such as interest, feelings, and beliefs (Popham, 2004). Therefore, AAMSUS's second evaluation goal was to assess whether or not students maintained or increased interest in advanced science and mathematics courses throughout the program.

Academic Outcomes Assessed

Ohio Achievement Tests (OATs)

In the state of Ohio, students in grades 3—8 take OATs each spring over the content areas designated for their grade level. Prior to the 2004—05 school year, valid assessments of student achievement did not exist in the state of Ohio as the old proficiency tests did not align well with what was actually taught in the classrooms. OATs were created with the purpose of rectifying this assessment problem by testing the state content standards that are taught by Ohio publically funded schools. Thus, content validity is high as the OATs were created based on blueprints of the state content standards, which Ohio public school teachers are required to teach in their classrooms.

All OATs measure five levels of student achievement. From highest to lowest they are: Advanced, Accelerated, Proficient, Basic, and Limited. Regardless of the grade level or content area, a scaled score of 400 is the lowest OAT Proficient score needed for passing. Individual tests, however, vary in their ranges for descriptive statistics (e.g., mean, standard deviation, reliability, etc.). Further, achievement level cut-score points for Advanced, Accelerated, Basic, and Limited levels vary slightly between content areas and grade levels, but remain the same within each content area for a single grade level from year to year. Therefore, while OATs are a valid measure of student learning in the state of Ohio, they cannot be used to assess growth over time because they are not vertically equated with all scaled scores meaning the same across grade levels or content areas.

OAT Results for AAMSUS Students versus Comparison Group. Unfortunately, students were only assessed in reading and writing during their fourth-grade school year, so we do not have mathematics or science results for their first year in the AAMSUS program. However, Mathematics OAT assessment began in the fifth grade and was done each spring through the eighth grade. The only Science OAT this group of students took was in the eighth grade. To compare AAMSUS student OAT achievement against a comparison group, students in the program were matched with non-AAMSUS students from the same school district based on grade level, gender, ethnicity, and third-grade Reading OAT scores (the only OAT given that year).

Mathematics OAT comparison results indicated AAMSUS and non-AAMSUS students were statistically similar in the fifth grade. However, in grades 6—8 AAMSUS students scored statistically better compared to their matched non-AAMSUS students on Mathematics OATs. Additionally, AAMSUS students scored statistically better on the eighth-grade Science OATs when compared to their matched non-AAMSUS students. It is important to note that when conducting these independent samples *t*-tests, we considered any test statistic resulting in a *p*-value < 0.10 as being statistically significant, since we were dealing with such a small sample size and looking for changes in a population that is very difficult to impact. See table 4.2 for specific statistics.

Practical importance of these results showed in the fifth-grade year (second year of the program). Both AAMSUS and non-AAMSUS students were on average not scoring at the proficient level in mathematics. However, from sixth to eighth grade the AAMSUS group consistently averaged above proficient in mathematics while the non-AAMSUS group continued to perform below proficient with the exception of their sixth-grade year. For science, although AAMSUS students scored significantly higher than non-AAMSUS students, both groups averaged below proficient on the eighth-grade Science OATs. While both groups did score below proficient in science, AAMSUS students were very close to averaging proficient (2.74 points away) while non-AAMSUS students averaged in the limited (lowest possible) category. Figure 4.1 depicts the practical significance of these results.

OAT Results for AAMSUS Students by Selection-Criteria Groups. Since there was great variation in how the AAMSUS students were selected for the program (high scores; high scores plus teacher recommendation; and teacher recommendation alone), we thought it was important to see if the program impacted the groups in different ways. Our findings showed that there were statistically significant differences between the three groups for all years they were tested on the Mathematics OATs, as well as a statistically significant difference between groups on the eighth-grade Science OATs. In general, when there were statistically significant differences between groups, students who were selected by their scores alone consistently scored statistically higher than those selected by teacher recommendation alone. Students selected by score alone scored statistically higher than students selected by score plus teacher recommendation in only fifth-grade mathematics. And,

Table 4.2 Independent Samples *t*-test Results for AAMSUS versus Matched Non-AAMSUS Students on OAT

OAT	AAMSUS Students		Non-AAMSUS Students			
	M	*SD*	*M*	*SD*	*Df*	*t*-statistic
Fifthth-Grade Mathematics	398.77	25.09	396.95	23.15	76	−0.33
Sixth-Grade Mathematics	421.92	26.37	405.54	28.21	72	−2.58***
Seventh-Grade Mathematics	406.36	22.46	393.24	22.17	64	−2.39***
Eighth-Grade Mathematics	408.45	24.12	382.94	73.33	60	−1.84**
Eighth-Grade Science	397.26	18.43	378.23	74.35	60	−1.38*

Note: $^{*}p < .10$, $^{**}p < .05$, $^{***}p < .01$

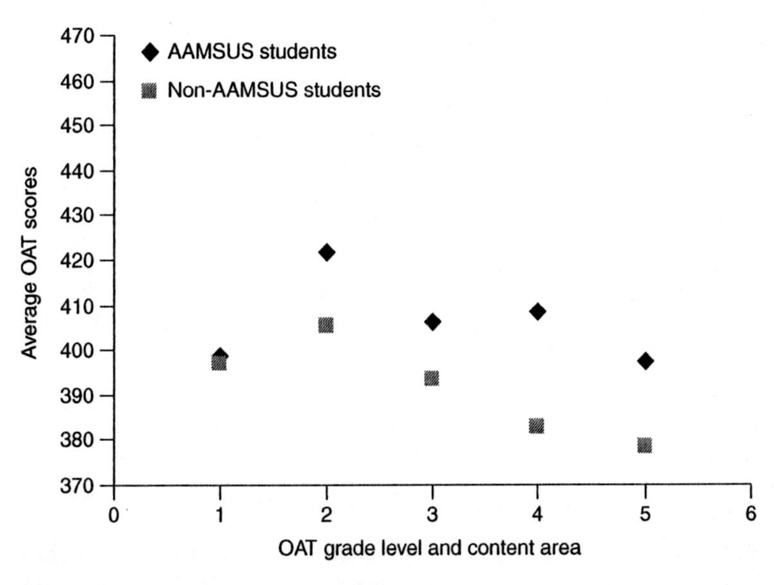

Figure 4.1 Practical Significance Showing Average OAT Scores Compared to Proficiency Levels for AAMSUS versus Non-AAMSUS Students

Table 4.3 Means and Standard Deviations for Three Identification Groups on Five OATs

	Scores Only		Teacher Recommendation Only		Scores + Teacher Recommendation	
OAT	*M*	*SD*	*M*	*SD*	*M*	*SD*
Fifth-Grade Mathematics	419.92	17.45	397.13	19.19	376.00	18.84
Sixth-Grade Mathematics	439.69	20.71	421.00	17.37	397.78	28.08
Seventh-Grade Mathematics	424.00	15.43	408.00	19.34	385.00	13.43
Eighth-Grade Mathematics	418.90	19.13	415.73	25.06	390.00	17.63
Eighth-Grade Science	403.20	14.06	400.18	20.13	388.10	18.45

students selected for AAMSUS by score plus teacher recommendation scored significantly higher than students selected by teacher recommendation alone on all years of the Mathematics OATs. Table 4.3 shows descriptive statistics and table 4.4 shows specific statistics for each year's one-way Analysis of Variance (ANOVA) group comparison.

In practical terms, students who were identified for the program by scores alone consistently scored above proficient on every OAT. Students identified for AAMSUS by scores plus teacher recommendation scored above proficient on every OAT except for fifth-grade Mathematics. While students identified for the program by teacher recommendation alone averaged below proficient on every OAT. Figure 4.2 depicts the practical significance of these findings.

Affective Outcomes Assessed

Modified Attitudes toward Science Inventory (mATSI)

AAMSUS students completed the mATSI each year for the first three years of the program (2005—07). The mATSI is a 25-item instrument that assesses the affective domains of engagement and attitude toward science. Students' perception of the science teacher, anxiety toward science, value of science in society, self-concept of science, and desire to do science

Table 4.4 One-Way ANOVA Results for Effects of Program Identification on OATs

OAT	*df*	*SS*	*MS*	*F*
Fifth-Grade Mathematics				
Between groups	2	11560.27	5780.13	16.83****
Within groups	36	12360.66	343.35	
Sixth-Grade Mathematics				
Between groups	2	9364.43	4682.22	10.16****
Within groups	34	15674.33	461.01	
Seventh-Grade Mathematics				
Between groups	2	8017.64	4008.82	14.81****
Within groups	30	8122.00	270.73	
Eighth-Grade Mathematics				
Between groups	2	5078.60	2539.30	5.75***
Within groups	30	12371.08	441.82	
Eighth-Grade Science				
Between groups	2	1285.80	642.90	2.02*
Within groups	28	8898.14	317.79	

Note: $^{*}p < .10$, $^{**}p < 0.05$, $^{***}p < .01$, $^{****}p < .001$

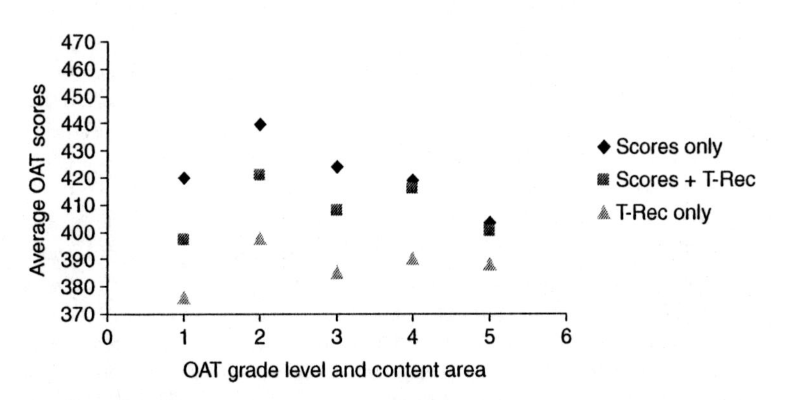

Figure 4.2 Practical Significance Showing Average OAT Scores Compared to Proficiency Levels for Program Identification Groups

are assessed on a five-point Likert-type scale (strongly agree, agree, neutral, disagree, strongly disagree). This instrument was modified to be used with urban middle-grade students and has a relatively high reported internal reliability (Cronbach's Alpha=0.70) when used with urban middle-grades students (Weinburgh and Steele, 2000).

mATSI Results for AAMSUS Students. In general, students attitudes toward science, as measured through the mATSI subscales (perception of the science teacher, anxiety toward science, value of science in society, self-concept of science, and desire to do science) did not statistically change over the three years the mATSI was administered to AAMSUS students (years one—three of the program). On an average, AAMSUS students were in general agreement with their beliefs over time on perceptions of the science teacher, value of science in society, self-concept of science, and desire to do science. Additionally, the average AAMSUS student indicated low levels of anxiety toward science as they reported an average disagreement to strong disagreement with science anxiety questions. Even though there were no statistically significant results found, these are positive findings for the AAMSUS program, since students entered the program with relatively high affect toward engagement and attitude toward science and maintained these positive attitudes through the first three years of the program.

Qualitative Assessment of Affect for the AAMSUS Program

A goal of AAMSUS was to increase attitudes and interest in science and mathematics as measured by retention in the project, identification of career interests in mathematics, science, and technology, and development of independence and expert ratings of performance in research and learning activities. We gathered data through a questionnaire adapted from the Midwest Academic Talent Search, interviews, and essays. We were looking to discover student interests for curricular-programming purposes and to see a growing connection between interest and career knowledge, such as choice of college major, course work needed for the career, and so forth. The majority of AAMSUS students expressed some interest in science and mathematics, but for only a few was it high interest. Knowledge of mathematics and science was rated as important, with mathematics somewhat more highly valued than science (80 percent versus 70 percent). When we asked questions about their interests, future college major, career goals, and possible occupations, the linkages were weak. In student responses the connection between personal interest, future schooling, and occupation was not closely related. Students stated interests in a variety of careers in the arts, engineering, medical science, civil service, humanities, sports, and education. Although careers in science and related fields were topics over the five years and even with focused courses on careers the last year, students did not make the connections we had hoped for. Among all the students, the more academic-oriented students exhibited the strongest sense of what they needed for a career, and parental interest was likely a stronger influence that the program.

Exit interviews conducted on the bus returning from the culminating field trip to Washington, D.C., were very mixed in their sense of their future and their understanding of what is ahead of them for further preparation. Incomplete would be the best descriptor. The students fell into three rough groupings. One group had a clearer view and direction, and it included the academic students. An equal number, comprising the second group, had an unrealistic notion of career linked to dreams of glory in professional sports or in pop culture. A third group had a direction that corresponded to their parent/guardians' occupations, which were semiskilled and/or public service careers. For example, one boy said, "I see myself in construction; it is in my family," and a girl declared, "I will be a bailiff like grandfather." AAMSUS was not successful in making inroads to students' naïve notions of career. The distance between middle school and career was a gap we were unable to close.

As part of the students' writing class in the final summer of AAMSUS, students composed essays on their five years over four class periods. Guidelines for the essay were that two paragraphs were to be written each day. Each paragraph was to be focused on topics such as field trips, science, computer, mathematics, other classes, and future goals. Students said that they learned mathematics and science from the field trips, and that they had fun. The overwhelming favorite field trip was the trip to Cedar Point. Marylou wrote, "We go on field trips to get more learning experience and to entertain us. [I remember] the fun we have when we are at the field trip. Many things I learned that you can have fun while still being educated. I look forward to everything on the DC trip."

Students remembered many lab experiments. Individual students mentioned experiments with DNA, forces, model roller coasters, pulleys and levers, owl pellets, structures, liquid nitrogen, bacteria, zero gravity, electricity, and Newton's laws. In general, students believed that the AAMSUS classes were superior to the science classes at their local schools. Individual students described the AAMSUS classes as very interactive, more fun than school, more challenging than school, and covering more concepts than school. Unlike school science class, AAMSUS classes did not require students to use textbooks or copy notes off of the chalkboard. Two students described AAMSUS classes as helping them to think scientifically and learn to develop and test hypotheses. However, a very small minority said AAMSUS science classes were boring and repetitive of what they learned in school.

Many students remembered computer class as their favorite course. They described computer class as fun and easy. A few students had their first experience using Microsoft Office as a result of the AAMSUS program. Some described learning more in AAMSUS than in their school computer classes, because computers in their school were used to practice typing or standardized testing. Students recalled working with PowerPoint, Google SketchUp, programming computer games using the SCRATCH language, using PhotoShop, doing internet research, and making movies. Two students reported that they practiced the skills they learned in class at home. There were no negative comments made about computer classes.

Most students found mathematics classes at AAMSUS to be of moderate difficulty: challenging, but not too challenging. A few students described the classes as very difficult; three students described the classes as easy. Yet some other students described the mathematics classes as very advanced and as fun. A few students also disliked mathematics and reported these classes were boring and repetitive of mathematics at school. Two students wrote that they learned nothing in their AAMSUS mathematics classes. Overall, students recalled learning algebra, trigonometry, problem solving, sequences, golden ratio, Fibonacci numbers, and the Pythagorean Theorem. DyJuan wrote, "I liked doing the mathematics problem we are doing in [the advanced] class. I learned many things I had not learned before. [The class] was hard."

Another indicator of student attitude was retention in the AAMSUS project. Our record is poor. Almost 40 percent of students were retained through all five years of the program. Most of the loss occurred in the first three years with the repeated school closings and reallocation of teachers and students. Our exit interviews indicated that retention was neither connected to identification category nor to the curriculum. In terms of category, all program selection groups remained in about the same proportion. Furthermore, it appeared that familial economic issues, mainly looking for work and movement from home to home, accounted for the greatest loss of connection to AAMSUS.

Lessons Learned and Practical Implications

As AAMSUS program developers, content specialist teachers, and a program evaluator, we learned a great deal about implementing a nonrequired accelerated mathematics and science program for at-risk, urban middle-grades students. Not all of the lessons learned, as illustrated above, were pleasant or expected. Yet even with teacher, student/family, and logistical challenges AAMSUS still had some victories to share as our students consistently outperformed matched non-AAMSUS students in state proficiency tests. Based on our experiences over the course of the five-year program, we offer the following practical suggestions for anyone attempting to conduct a program similar in nature.

- **Teachers**—Using teachers from the students' home schools as instructional leaders and mentors seemed promising for building a necessary support network for at-risk urban students. However, we found out almost immediately in the program that early middle-grades teachers (although certified to teach K-8) did not feel confident teaching higher middle-grades mathematics and science content, or even their own-grade-level mathematics and science content if asked to teach through nontraditional methods that they were unaccustomed to using (differentiated or project-based instruction). For future programs, we recommend adopting the Teacher Liaison role and the Teacher Content Specialist models that we ended up with in the last three years of the program. This was effective for increasing student learning in that content specialists with educational backgrounds were confident to deliver high-quality instruction through a wide variety of instructional methods. And, teacher liaisons acted as behavioral and instructional support, providing structure and consistency in the program and during transportation. Teacher liaisons and teacher content specialists each played their own role well. No one group was responsible for both roles allowing each to specialize and focus on what they did best, mentor or teach students respectively.
- **Curriculum**—An enriched curriculum using acceleration is an instructional approach that warrants more attention. Inner-city students with high potential are able to move at a faster rate into more challenging mathematics and science content when those opportunities are provided. The whole group, move at the rate of the average students, in the classroom works against increasing achievement of these more capable learners. Inner-city children will become higher achievers when we enable them to experience more accelerated curriculum. While our students had difficulty making this adjustment initially, with the move to content specialist teachers and repeated encouragement, the majority of our AAMSUS students successfully made this switch by the end of the program.
- **Students**—Middle-grades students are amazingly complex, changing from children to adolescents. Social interactions often outweigh educational opportunities regardless of the academic promise the student possesses. For these reasons, it is imperative to provide

engaging content, and that allows for social interaction in nonrequired academically focused programs with middle graders. It is also important to have consistent adult supervision and leadership throughout the program. The urban middle-grades children participating in AAMSUS did not immediately embrace new content specialist teachers in the first couple of weeks or even months. Instead, it took the better part of a year working with these students before a relationship formed where students began to respect and trust content specialist teachers to "be there for them" at each session. Much time was invested in developing productive relationships between the students and the content specialist teachers as the teachers had to build a sense of community and communication first (Ullucci, 2009). We firmly believe that many of our students' somewhat turbulent personal and school experiences may have nurtured these cautious relationship-forming feelings (Pellino, 2007). Thus, having the same content specialist teachers involved in the program from day one would most likely have produced greater impact on student achievement as students would have been willing to engage and participate more fully in program content earlier.

- **Parental Figures**—Getting parental figures of middle-grades urban students to buy into the program is the first step in retaining students. While every AAMSUS parent wanted the best for their child and saw that AAMSUS would be beneficial for their child's future, they did not always support these feelings through their actions. We believe that they were unprepared to support their children to be in a program such as AAMSUS, and we did not do the best job of helping them sustain their child in the program. Providing initial counseling for parents at the very first informational meeting about the educational and social experiences was an important first step, but continual counseling is key to retention. Topics such as discussing the importance of mathematics and science to their child's future careers, handling the students' discomfort with the academic challenges of the program, dealing with growing independence in student interests, and so forth should be the basis of continual counseling.

The AAMSUS project shows that changing achievement in science and mathematics in UPSS is a complex challenge, but it is possible to make a small difference. Much of what happens in students' lives outside of school is out of our control, yet we have the power over what happens in our classrooms. Using acceleration is one such approach we can control by which we can raise achievement. Essentially, this is done by providing opportunities for children by allowing their teachers to move students ahead of age-grade-subject conventions when children demonstrate that they are ready for such advanced experiences.

The present whole group, stay on grade level, skill and drill method of instruction as practiced in the classrooms of this study is in opposition to acceleration. Achievement scores of students in the control groups were lower than the AAMSUS students. Yet within that standard approach, assessment data are gathered that reveal some children have mastered the material and are ready to move on. These data should be used to implement acceleration for students who are ready. Children will learn more when they are given opportunity and support to achieve, as we've seen here with our AAMSUS students when compared to the non-AAMSUS comparison group. To make this happen we need teachers who are knowledgeable about advanced content and differentiated teaching strategies, and we need schools to support teachers who are willing to do so. What are we waiting for?

Chapter 5

The Right Tools for the Job: What's the Role of Literacy in STEM Initiatives?

Holly Johnson and Patricia Watson

Many people consider technology as a tool, but it is actually content. It requires knowledge of the math and science behind it, as well as knowledge of how it is applied.

–*A. Dean Fontenot, engineering educator*

Learning science is a human endeavor. Scientific knowledge is an accumulation of the human pursuit, and it's the same as history or writing novels as that accumulation of human experience.

–*Helen Meyer, science educator*

Science, Technology, Engineering, and Mathematics (STEM) disciplines have the potential to advance secondary school knowledge and education beyond the borders of what is currently known or was believed could be possible. Projects that provide opportunities for interdisciplinary work that cross all the STEM disciplines are especially fruitful, yet knowledge is often taught in isolation, without consultation of how multiple disciplines are needed to create new understandings of the world. There are strategies and skills that might be utilized for bringing these disciplines together, especially in respect to inquiry and discovery. "Tools" of discovery and inquiry are frequently referred to as literacy tools that include reading, writing, viewing, and visually representing. Tools of communication also include speaking and listening. Regardless of the STEM discipline, these literacy tools can hold the STEM disciplines together within particular initiatives.

In this chapter, we discuss our current understandings about disciplinary literacy, its relationship to the STEM disciplines, and its capacity for creating interdisciplinary inquiry and discovery. Our inquiry into the connection of literacy and STEM initiatives involved working with disciplinary experts whereby we discovered the dominant questions asked in the field, the texts they would use for exploration and inquiry, and the language utilized by the field for reporting what is important to know or learn. Through these explorations we found that there are particular literacy tools that could be determined to be "the right tools of the job" in learning and knowing what it means to think about math, science, technology, and engineering. We share this knowledge along with what schools and secondary teachers might do to encourage students to think within the STEM disciplines.

Disciplinary Literacy and STEM

STEM initiatives often miss the need for literacy skills and strategies in their quests for knowledge advancement. Yet, they do require literacies conducive to what it means to "know" in the STEM disciplines. Generic literacy skills and strategies are often inappropriate, and thus remains the question of which literacy tools are best for a project. To address which literacy tools are most conducive for particular initiatives, the field of disciplinary literacy has taken central stage in the larger field of literacy.

Disciplinary literacy has a long history rooted in the paradigm of applying cognitive strategies to content-specific texts (Moje, 2008). Developing from Gray's (1925) work with study skills at the beginning of the twentieth century and continuing through Harold Herber's (1970) work, the idea of teaching cognitive strategies for making sense of text dominates content-area literacy textbooks. Recently, however, theory in content-area literacy has begun to focus on adolescent literacy, including issues of culture, social interaction, technology, and diversity. Researchers and theorists have encouraged those involved with content-area literacy instruction to adopt a disciplinary approach, which is a more complex view of literacy instruction that addresses the literacy demands specific to content areas.

This approach is based in the belief that deep knowledge of a discipline is best acquired by engaging in the literate habits valued and used by experts in that discipline (Moje, 2008; McConachie, Petrosky, and Resnick, 2009; Lee and Spratley, 2010). Through reading, writing, and thinking in ways common to the discipline, students deepen their knowledge and understanding of disciplinary content. This paradigmatic shift presents enormous challenges to those of us engaged in the preparation of teachers.

Stemming from an examination of the difficulty of infusing literacy into secondary school instruction, O'Brien, Stewart, and Moje (1995) indicated a need for educators to "rethink the philosophy, epistemology, and goals that underpin content literacy research and teaching" (p. 442). They suggested that the strategy instruction prevalent in content-area research and teaching needed to consider the contextual constraints of content classrooms. The professional development of teachers would then move away from providing content-area teachers with strategies without knowledge of the particular discourse communities in which they would be used. Draper et al., (2005) described this disjuncture between content-area instruction and literacy instruction as a "dualism," and their contention was that teachers must explicitly instruct students about how the texts in their disciplines are created and used. Draper's (2002) work with her content-area colleagues yielded the understanding that some of the content-area strategies suggested by literacy educators were actually contradictory to the needs of some disciplines and encouraged content teachers to engage in literacy instruction that did not support learning in the discipline.

Many of the strategies advocated in content-area literacy textbooks today were based in the work of Harold Herber (1970), who introduced the concept of providing instruction for the reading of expository texts in content-area classes. A variety of generic cognitive strategies emerged, aimed at the comprehension of texts from all disciplines. Other research, however, has shown that some generic strategies have the potential for supporting readers in comprehending many different types of texts (Heller and Greenleaf, 2007; Moje, 2008). The current challenge for teachers and teacher educators is to identify which strategies have merit and which should be discarded, as well as to develop their own knowledge of the roles texts and literacy play in the disciplinary subject areas in secondary schools.

The definition of literacy, in general terms, has also changed with the passage of time. At the time our country was founded, people who could sign their names were considered literate. Not until the early twentieth century did employers expect their employees to read previously unseen text. Since 1980s, we have witnessed a dramatic increase in the need for employees who could use literacy skills for creative and critical thinking

(Braunger and Lewis, 2006). The term literacy, historically, has referred to the ability to comprehend and write text. The term, now, however, extends to literacies of the STEM disciplines. Looking toward the twenty-first century, Secretary of Education Richard W. Riley established a National Commission on Mathematics and Science Teaching for the 21st Century. Chaired by astronaut and former senator John Glenn, the commission was charged with creating an action strategy to improve the quality of teaching in mathematics and science at all grades nationwide. The commission's report emphasized the need for an educated citizenry, literate in science and mathematics. The report stated,

> Mathematics and science have become so pervasive in daily life that we tend to overlook them. Literacy in these areas affects the ability to understand weather and stock reports, develop a personal financial plan, or understand a doctor's advice. Taking advantage of mathematical and scientific information does not generally require an expert's grasp of those disciplines. But it does require a distinctive approach to analyzing information. We *all* have to be able to make accurate observations, develop conjectures, and test hypotheses—in short; we have to be familiar with a scientific approach (p. 14).

The definition of literacy in the STEM fields is shifting and changing as these fields define themselves in relation to K-12 education and the expectations for career and college readiness. Recently, Chae, Purzer, and Cardella (2010) examined the definition of STEM literacy, seeking commonalities and differences between the fields of engineering, technology, science, and mathematics. Their review of the popular publications of the fields found that the definitions of literacy in the STEM fields shared a common emphasis on the ability to problem solve. Similarly, the STEM fields share an emphasis on the use of models when organizing and sharing concepts and ideas. Their review also revealed an emphasis on societal impact within the STEM fields. The STEM fields have developed around the needs for the improvement and development of our society.

The task force of the National Governor's Association's *Innovation America* initiative sought to strengthen the competitive position of the United States in a global economy, in part through improving STEM education. In their report (2007), the task force emphasized design and problem solving in an interdisciplinary setting, stressing that STEM literacy bridges the four disciplines of science, math, engineering, and technology. The report stated that classrooms that focus on STEM literacy should shift "students away from learning discrete bits and pieces of phenomenon and rote procedures and toward having investigating and questioning the interrelated facets of the world" (p. 7).

The current challenge for teachers and teacher educators then is to identify which strategies can best address the STEM disciplines, and which should be discarded. Furthermore, teachers and teacher educators will need to develop knowledge of the roles texts and literacy play in the STEM disciplinary subject areas in secondary schools.

Discovering Appropriate Literacy Tools

To address the literacy requirements of particular STEM disciplines, we had to understand what it meant to be literate in those content areas. Furthermore, we had to explore the types of texts used, the language of the discipline, and the important ways of thinking within each discipline. In essence, we had to know the discourses of the various fields. Although a few researchers began to describe and delineate the literacy practices of different disciplines (Donahue, 2003; Draper, 2008; Shanahan and Shanahan, 2008), we recognized that there was still more to know and to understand. Guiding our inquiry into the STEM disciplines were these questions:

- What does it mean to be literate in particular disciplines?
- How can teachers begin to include appropriate literacy skills and strategies into the STEM disciplines?

To find the answers to our questions, we met with arts and sciences colleagues from our respective universities, as well as those from engineering. They were also partners with us in delivering content appropriate for teacher candidates. We worked with two mathematicians (Smith and McSwiggen), one colleague in chemistry (Breiner), three people who work extensively with engineering and technology (Fontenot, Hambright, and Harris), and one science educator (Meyer). Our collaboration with our colleagues is an example of the kind of work that is needed to develop the field of disciplinary literacy and was framed within the context of informing teacher preparation. Interviews conducted with our selected colleagues in the STEM disciplines consisted of ten questions. The protocol addressed the concept of literacy in the discipline and what it means to be literate. As part of this chapter, we often use not only quotes, but also summations of what our colleagues expressed in respect to their fields of study.

Prior to addressing key questions through interviews with disciplinary colleagues, we also collected data from the following sources: textbooks developed for content-area literacy courses in teacher preparation, informal discussions with teacher candidates, major academic journals and disciplinary websites, and our lived experiences as content/disciplinary literacy instructors in teacher-preparation programs. We teach in middle grades or secondary education programs for candidates interested in teaching in grades 4–12 and have taught content-area literacy for students in these programs.

As part of our exploration of the STEM disciplines, we selected 12 current textbooks that are used to teach content-area literacy. The texts were divided into pedagogical knowledge and content knowledge and attended to "how to teach?" and "what to teach?" respectively. We noted the disciplines addressed within each text as well as the major themes, which included addressing the standards, the concept of multiple literacies, and the types of strategies suggested.

For this project we also reflected on our own experiences in the university classroom with teacher candidates. As content-area literacy instructors, we offer courses that address literacy across the curriculum at the undergraduate and graduate level. At the graduate level, where our courses are electives, students took our courses for purposes such as expanding their own knowledge base or as part of their plan of study. At the undergraduate level, students were required to take our courses, and they often challenged us with questions about why they needed to know about literacy if they are going to teach math, science, or other disciplines related to STEM knowledge. They challenged, through their skepticism, the textbook or the types of strategies we believed our candidates could pursue in their 4–12 classrooms. And, as a final source of data, we consulted disciplinary websites and journals of professional organizations to gain a better understanding of the particular STEM field, and to substantiate our thinking about the various disciplines as well as add to the advice our disciplinary colleagues shared with us.

What we learned from our inquiry we placed into categories such as "major understandings and practices of the discipline," and "being literate in the discipline." These categories were broken down into subcategories addressing the texts used in the discipline, the types of strategies used to become literate in the discipline, and the types of questions central to the discipline.

The Right Tools for the Job

Finding the "right tools for the job," whether in STEM or other disciplines must address the major understandings of the field—what it is that discipline does? what questions it asks? and what counts as knowledge? Discussing literacy in respect to any one discipline, however, requires consideration of further questions. As literacy educators, we wanted to know about the language of the field, the texts used within the field, and the common practices of the field that functioned as communicative measures for knowing the discipline. In the following sections, we discuss the major understandings of the STEM fields as related to us by disciplinary experts. We also discuss the language and texts of particular STEM fields, and finally some of the literacy practices within each field. As we venture into this discussion, it is important for us to note that while we may discuss "mathematics" or "science," we are well aware that there are subfields within each of the STEM disciplines that may have within-group differences that will not be discussed in detail here. We further note that what we discovered with our colleagues may not hold across all fields, subfields, or disciplinary experts. This chapter is devoted to the need of discovering what teachers might do in respect to the literacies useful within particular STEM domains, and there is still much to discover in respect to STEM disciplines and their literacies.

Major Understandings of the Disciplines

In this section, we discuss how the STEM disciplines define themselves and what those in the field do. This section also addresses the major questions asked in each of the fields. The major understandings in the STEM disciplines are varied, but when looked at closely, most focus on patterns. The questions asked in each field are specific and inquiry based, which allow for further construction of knowledge and discovery.

In the sciences of biology and chemistry, biology is "concerned with causation and we represent this in models using theories that will elaborate causation." Science as a general field examines patterns and uses those patterns to create better understandings about the world, about individuals in the world, and about cultural ramifications in respect to the individual in the physical world. In addition to causation, there is also prediction, which is a result of observed changes to theoretical models or generalizations that allow for prediction of changes in the future. Science utilizes models and theories in respect to patterns, with models including experimental designs, which could be diagrammatical. The sciences also utilize mathematical models involving symbols, graphs, and numbers, and there are also theoretical models that guide the thinking in the field. According to Breiner, chemistry often seeks to understand what happens generally at the molecular level so as to extrapolate what is happening with particles to explain what is happening at the macroscopic level. Visual representations as well as statistical representations aid in these types of models, and as Meyer pointed out, "Going from the macroscopic world across many of the science disciplines, there are three-dimensional models that are almost artistic in nature."

Questions in science stem from causation. In chemistry, much of the knowledge concerns questions on a molecular level. Some of the current inquiry addresses the functions of genes and genomes at the molecular or submolecular levels. The questions revolve around reduction so as to investigate what is happening at the nano level, or to explore adaptation or life at the fringes that occur at the smallest levels. In biology, current questions revolve around life at the extremes, so biologists want to know about the limits of adaptation and extreme environments such as space or the deep sea. Exploring life at the extremes of our planet builds toward the question of could life exist in other solar systems. Both the science educator and the scientist agreed that what happens at the molecular level can have an impact at the macro level, which can lead to further questions and solutions to problems scientists are currently even exploring. There are a variety of questions that science addresses, and both of the participants repeated that different areas of science may be looking at other global questions not addressed by chemistry or biology.

According to Harris, who is a computer scientist working for a major technology firm, technology as a discipline is really about the generation of knowledge in search of solving problems, with problems being a wide and purposely unspecific term. Fontenot, looking at technology from an engineering perspective, also stressed the emphasis on knowledge in the pursuit of solving problems. She added that in the STEM disciplines, technology is both a set of existing tools that are used in solving problems as well as the innovations and creations developed by engineers in search of new tools. Harris emphasized that in the technology fields, inquiries seek patterns related to how things work and why they work. This general inquiry is then narrowed down. The questions asked are why is this occurring, how can we accomplish this task, and why is this important? Some individuals choose to look at these questions from the software or data side of technology, while others choose to look at the hardware side of computers or other technological tools. Eventually they must come together in pursuit of answers to specific problems, but generally their inquiries begin from different points of entry. However, Harris stressed, they all problem solve and seek repeated methods that can be applied in the pursuit of solving a variety of task specific problems. These repeatable methods are important in all aspects of technology.

The field of mathematics typically addresses pattern recognition and generalizing, as well as mathematical models represented in one of three ways: numerically, symbolically, or graphically. Proof is the foundational knowledge base. Mathematician Patrick McSwiggen stated, "Ultimately, [what] the mathematician wants to do is prove something about [a mathematical] model." Mathematics, however, does not typically run simulations or create computer models to then assert that there is an answer. McSwiggen says that mathematicians would venture, "That's a nice example, now prove it." Tara Smith defined mathematics' conceptual framework that "lets us understand 'number'; that lets us understand logical proof."

Many mathematical questions and inquiries about patterns involve "what if?" thinking. When mathematicians are given a mathematical model, they ask questions about removing a postulate or adding a component to see what will happen to the model. Smith added, "Applied math is much more 'let's see what we can come up with.' We've identified these patterns; now let's see if we come up with a mathematical model that would generate those patterns." Mathematicians will often ask related questions to the "what if?" that address the veracity of a model or pattern that involve questions about the model fitting what can be observed in the real world or whether or not the model will always hold. In other words, mathematicians want to prove that whatever is observed in the model always occurs. Mathematicians also ask what will happen if something is added or removed to change the model's pattern. For Smith and McSwiggen, mathematics is not about questions of doing equations, but rather more interested in generalizing a particular pattern. Smith suggested that most mathematicians are about discovery of knowledge and theorems within Platonic realities.

When comparing the questions, definitions, and major understandings of the STEM disciplines, we note the use of patterns across fields, but would suggest the content and purposes of those patterns are qualitatively different. One field is about finding and describing patterns to make predictions about the physical world, another is about testing or proving the patterns and under what conditions. The mathematicians observe the real world and create patterns through numeric, symbolic, and graphical representations, and then attempt to prove those patterns as a way to establish mathematical truths. Biologists use their observed patterns to make predictions about living things in the physical world. Technology as a discipline seeks repeatable methods that can be applied in the pursuit of improving a process, making a task more efficient, or creating a tool to achieve a purpose. Technology and engineering (the T and the E in STEM), are really the disciplines where the patterns that emerge from inquiries in science and math are applied in practical ways in the pursuit of solving human problems.

Dominant characteristics in the STEM fields can also be viewed by precision in juxtaposition. Each of the STEM discipline has a need for precision and will seek assurances of that precision through language. The "what if?" of math relies on manipulating variables in juxtaposition, while science and technology rely on how particular elements are juxtaposed to create a pattern that either represents living things in relation to the physical world, the way particles work at the molecular level, or are used in logically thinking through a problem to find a technological solution to a human need.

Literacy in STEM Disciplines

Through our work with disciplinary experts in the disparate STEM fields, we sought to identify the specific kinds of knowledge needed to comprehend texts and construct knowledge in those fields. While we sought to understand our disciplinary colleagues' ways of reading and thinking, we also presented them with cognitive tools and strategies commonly taught in content-area literacy courses, with the goal of identifying those that would be most useful and relevant to novices in these disciplines.

Our colleagues in science suggested that to be literate in science, there is foundational knowledge to know and then to build upon. There is a need to understand the organizing structures of many of the sciences so those who "do" science will know how to build upon that foundation. Being literate in chemistry includes having the ability to obtain knowledge through multiple means. As Breiner noted, "That could be piecing together knowledge from other disciplines, or again, knowing how a scientist works and acquiring that knowledge. It is understanding when [one has] the answer to their question and when [they don't]." For biology, being literate is being able to work with and within theories, to organize around theories, and to find patterns that fit or don't fit the theories. Meyer asserted, "How to derive patterns. To be literate in biology you have to know the patterns that are most prevalent and how to look for the domain knowledge, as well as how they interact with one another." Thus, to be literate in the sciences, students will have to know the foundational knowledge, how it relates to theories, and how systems interact. To be literate is realizing that isolated bits of knowledge fit within a pattern, an organization, a system, and how these all interact. Scientists build on the conceptual, always looking for relationships.

In technology, being literate assumes a set of understandings that are used to solve personal and societal problems through either the use of or development of technological tools and processes. Fontenot and Hambright presented the example of a group of engineering students who were working together to design solar panels to provide sustainable energy for a public arts center in the community. First, an inclination toward relentless problem solving is required. Then, they explained, the students need to have the math and science background necessary to understand solar energy, as well as the skills necessary to design and use new tools and new technologies. In addition, there is also an element of critical literacy required in that the students must be able to examine and understand the societal implications of their work. Students of technology must learn to consider the outcomes and results of innovations as citizens of society. They must appreciate that the development of new technologies involves trade-offs and that there must be a balance between the benefits and risks.

Since the 1980s, when personal computers became widely available for school use, there has been a continuous emphasis on technology education spanning all grades. Early in the twentieth century, technology was taught in the context of industrial arts (shop), and then moved to instruction in computer skills like keyboarding, word processing, spreadsheets, and basic programming. In recent years, more schools have tried to include design projects. Even today, students studying technology have little experience with open-ended problems. Harris stressed that the ability to approach ill-defined and open-ended problems is critical in the technology workplace of today. He said that "not being able to search for solutions, not being able to problem solve and figure things out on your own, that would be considered illiteracy [in my workplace]." He stressed the need for ambiguity, or learning situations in which there is not a single right answer.

Technology, when considered as one of the interrelated STEM disciplines, requires practical and theoretical knowledge of tools and processes. It also requires ways of thinking that involve asking, questioning, seeking information, and participating in decision making. These attributes are in addition to the basic capabilities of hands-on skills with computers and other technological tools. Fontenot and Hambright stressed that Project-Based Learning (PBL) is one of the most useful frameworks for learning and applying technology-related skills and ways of thinking. PBL (Buck Institute, 2010) is organized around a driving question or challenge. This instructional framework requires students to engage in inquiry, collaboration, and communication, as they apply new knowledge and skills to complex problems.

Mathematicians observe, theorize, and prove arguments. To become a mathematician, the most important attribute a person might have is the proclivity to ask the right kind of questions, which Smith asserted is "the biggest literacy tool." Mathematicians must think in logical, sequential, and structured ways. They must be organized, and they must use precise language in their efforts to expand the thinking in the field. The common arithmetic taught in grades P-8 forms a foundation for mathematics, and typically what happens in those grade levels is the use of example and observation of patterns. Moving beyond example and observation to the creation of valid proofs is the hallmark practice of mathematicians.

Becoming a mathematician is developmental and can begin at an early age if given the appropriate tools and guidance. Math students, then, need the opportunities to play with or work through the knowledge they are learning. What mathematicians do is converse, and K-12 students can be apprenticed through their own inquiries and through invitations such as math circles that disrupt the resistance "of laying themselves out there on the line and [finding that] maybe they're wrong" (McSwiggen). Thus, talk in math circles, where students can make concrete what they are forming in their heads, and thinking aloud, where the formulation of their thinking takes place, are two of the most powerful literacy strategies teachers can use in mathematics classrooms.

Reading math, and not just math examples, would also allow students to begin utilizing the grammar of the field, as would the play with and creation of logic puzzles. In addition, we would also suggest students read classic mathematical texts, and then write their own so they can begin to utilize the grammar of mathematics. Comparing English and mathematical phrases in a T-Chart would also be a useful literacy tool for young people developing as mathematicians. And, the writing of simple proofs, parsing examples of proofs, and the discussion of sequence as they marked their charts or made Venn diagrams would also help students to articulate their mathematical reasoning.

According to McSwiggen and Smith, mathematical literacy involves the functional "number and graph sense" but extends to interpretation of graphs and the connections between numerical and graphical representations. To be literate in the field of mathematics, however, requires the ability to be conversant in your particular field of mathematics. It involves being well-read, or the ability to discuss the major results in the field and to read a research paper in your field. McSwiggen asserted that mathematics involves "this developed skill of being able to think through an argument... [which is] what one would like to have addressed when children are being educated."

Interestingly, regardless of STEM discipline, one of the most powerful literacy tools utilized is metaphorical thinking. Yet, to create metaphor, students need to understand the disciplines and their uses of models and patterns. Note taking and note making are the primary literacy tools K-12 students could use for identifying intertextual patterns as well as for the collection and organization of information in order to make intertextual connections that are crucial for metaphorical thinking. Visual literacy is another crucial tool in the STEM world. Learning to "read" all sources of data is important in the work of those within the field, and grids and questioning strategies like those provided by Ogle, Klem, and McBride (2007) prompt students to examine the purpose, source, point of view, and impact of primary source documents and images and can lead to the kind of metaphorical thinking necessary for deepening their content knowledge.

Making the content of science, technology, and mathematics concrete in students' heads means content-area literacy educators will need to work more closely with disciplinary experts like those in our project, so all those involved in educating for STEM knowledge can become more concrete in our knowledge of each other's disciplinary strengths. With the emphasis on juxtaposition in respect to patterns and processes, being conversant in their fields, and the ability to make connections regardless of STEM field, there might be an assumption that the texts they use would be similar. Yet, a simple examination of the content texts used to teach these disciplines in K-12 classrooms reveals vast differences. What are the texts commonly used within the STEM disciplines? What about the use of language within the fields?

"Texts" of the Discipline

We embrace Smagorinsky's (2001) definition of text, which "refers to any configuration of signs that provide a potential for meaning" (p. 137).

When considering the texts of science, there is the whole world at the macroscopic and microscopic levels. Many of the texts are mathematical in nature with biology focusing on probability, the statistical, and the large scale. Over time, scientists have tried to develop mathematical models to explain theoretical concepts, thus the significant mathematical text—and procedural text—that is involved in scientific inquiry. Observation is "a huge thing" in the sciences, according to Breiner. Thus, texts can be notes of a system that is observed, a detailed list of what is seen or perhaps not seen, as well as organized data and evidence from the observed or theoretical. With any text in science, there is always the pattern to observe, organize, or systematize. Taxonomies are also the texts of science. Thus, there is always further evidence to read and/or write, and thus knowing multiple symbol systems in English, Greek, Latin, or other languages is important in accessing the texts of the disciplines. In many ways, everything is a text within one science or another.

The texts of technology include a wide variety of sources and types of media. The pursuit of innovations and solutions to problems may require the application of knowledge and concepts from the related disciplines of science and math. Fontenot, Hambright, and Harris all stressed that the ability to read and comprehend traditional texts in order to acquire background or foundational information related to the history or theory of innovations is a critical ability. Harris explained that, in the world of technology, there is a hierarchy of theoretical and procedural texts, with the higher levels assuming knowledge and understandings presented in texts that are lower in the hierarchy. He described how it is often necessary to seek out texts that are lower or higher on the hierarchy in order to locate the information appropriate to the reader's level of understanding. Fontenot also stressed the importance of applying existing knowledge, often acquired through traditional texts, in the generation of new ideas, products, and processes. Included in these traditional texts are proposals, reports, description/specification papers, and academic publications. Fontenot and Hambright stressed the importance of learning to both comprehend and write these traditional texts. They stressed that, while engineers and others in technology fields must read and write texts that are very specific to their disciplines, learning to comprehend and think critically in general terms establishes a base of knowledge for later development of more specific disciplinary literacies. Harris agreed that "reading and writing... the language of communication in the traditional sense of literacy, is vital."

Using digital media and environments to communicate and work collaboratively is also critical to success in technological pursuits. Harris explained that many of the problems or tasks he is assigned as a software designer require him to use a wide range of resources ranging from traditional texts, search engines like Google, manuals, standards documents, codes, and samples, user-generated content like forums and online tutorials. He said, "There is a wide variety of 'text' accessible that contains bits and pieces of information that may be useful in solving a problem. It may be a direct explanation of how to accomplish something, a code sample through which you learn a built-in function or a book where you learn general practices."

Harris also described two specific types of texts that require discipline-specific reading skills. He said that much of the knowledge used in solving problems with technology is represented as either a tutorial or as documentation. Tutorials are procedural texts, requiring the user to read and follow a set of specific instruction. Documentation, on the other hand, resembles a glossary with specific terms and functions listed, often in alphabetical order.

The wide variety of texts used in technological problem solving requires users to locate, organize, analyze, and use information from a variety of sources and media. This range of sources also requires the ability to select, from available resources, those texts and tools that are most appropriate to specific questions and tasks.

The texts noted by the mathematicians in this study include "the incredible amount of knowledge in their heads," which comes from reading journal articles and other written works in their field or related fields, lectures and presentations of ideas from others, and from "just plain thinking about what they've observed and what they've read and internalizing it." According to Smith, this enables mathematicians to "see connections across areas because they have this huge wealth of knowledge." Other texts include examples and proofs from others in the field along with the tools, techniques, and theorems others used to prove their arguments. Smith contended there is a "definite style to mathematical writing that takes time to become comfortable with. Students struggle with 'how do I read mathematical texts?'" She continued by stating, "So many of the textbooks now are just busy, busy examples—fluff." With this she further asserted that mathematics has its own language and grammar, which can more readily be seen in the classic texts of mathematician Dolciani, who wrote a series of *Mathematics Structure and Method* texts in the 1960s. Reading Dolciani's mathematical arguments begins to develop mathematical grammar and thinking in respect to proofs and arguments.

The Language of STEM

When we discuss language, we are addressing the vocabulary used, the sentence grammar of the discipline, and the way in which symbols may be used. The language of STEM is formulated around proofs, patterns, and communication. In some of the STEM fields, the language is specific but borrowed from multiple fields across the disciplines of social and physical sciences. The reason? Technology. The vocabulary of all the fields has changed as the terminology specific to the technologies they use changes or becomes more dominant. There are times when the technology becomes the common language of the field.

In science, the language can often seem exclusionary but has patterns within that language. The Greek and Latin within its taxonomical information address naming structures or organizational patterns. With all this "labeling" as Meyer calls it, scientists are organizing the knowledge base. Breiner suggested that while it would be important for all the sciences to have a common language as well as symbol system, often naming conventions have been ignored because other everyday vocabulary has been used to discuss compounds or chemical nomenclature. As he stated, "There are huge issues in that regard because the teacher can use one symbol, the text another, and on the internet still another, which is very confusing [to those learning]." Meyer asserted that with the oft-time confusing nature of language within the sciences, students generalize to the whole field and thus may not like science, "The language is intense, particular, and often we don't attend to the historical development of how the taxonomies occurred…so it's difficult to bridge that in a timely way. Students just see a lot of words that seem to be disorganized." Thus, both Breiner and Meyer believe that language would be best discussed after teaching scientific concepts. Working with the confusing and diverse language history is one way of allowing more students to learn the language that is so necessary to the diverse sciences.

Within today's world there also exists a wide variety of technologies. There are medical technologies, agricultural technologies, construction technologies, and the ubiquitous computer and information technologies. Harris explained that each of these technologies has its own very specific "terminology." This terminology is a vocabulary that is very specific and unique and is the language required for communicating about the specific technology. He explained that within his workplace there are really three types of language: the logical programming languages that are unique to the computer world like C++ or Java, the "terminology" that is specific to the design, development, and use of computer systems, and the traditional verbal and written language that facilitates communication. Harris stressed that the environment in which technologies are developed is collaborative. Fontenot confirmed that teams rather than individuals accomplish much of the work done in the engineering fields today. These collaborations require facility with communication through all the languages of the specific technology.

According to the International Society for Technology in Education (ISTE), much of the work with technology is about applying existing knowledge to generate new ideas, products, or processes (2007). Harris described his working environment as a "culture of collaboration." Many of these collaborations stretch beyond the immediate workplace as peers and experts share knowledge freely. Harris stressed that learning the logical language necessary for the efficient use of search engines is an invaluable skill as these search engines are often the first "text" used in locating user-generated content that is widely available through forums and other "sharing" mediums.

Fontenot also stressed the need for facility in the use of both traditional- and technology-specific languages. She said, "Engineers must have more than just technological skills. There is a pressing and continuing need for clear communications. Conventions are still important as engineers write specifications, communicate through email, and communicate knowledge to wide audiences."

According to both Fontenot and Harris, the specific "terminologies" or language within the field of technology are learned largely through a culture of apprenticeship and mentoring. Harris described a typical team meeting in his workplace, saying that if a person were to visit, it would be very much akin to sitting in on a meeting conducted in a foreign language. Some of the language would be comprehensible, but much would not. Much of this language is learned through context as team members read and search for information as part of the problem-solving process. However, discussions include a great amount of clarifying talk and thinking aloud as team members explain their thought processes to others. Fontenot also detailed the importance of being able to apply background knowledge while actively questioning in order to learn the language and concepts pertinent to the project at hand.

The language of math is different from English, but both are used to understand mathematics. The key for mathematics language is precision and careful definition. There is also a grammar that typically begins with "Let" as in "Let 'A' equal..." Apprentices of mathematics must learn the language of math, the syntax and grammar of proof, and learn how to read the classic texts, which utilize more mathematical language than current textbooks. Math language is highly structured and sequential. McSwiggen stated that it is difficult for many students to "grasp the distinction between example and proof." When asked, students are often able to give an example where something is true, but they cannot generate the reason *why* it is true. Smith gave another example where English and mathematics are convoluted:

> What's shocking to me is the struggle with the inclusive "or." When I tell my kids you can have ice cream or cookies, I don't mean both. But, mathematically, if I say "A or B" is true, it means at least one. It could be more! So, there are those kinds of things about the way the language is very precise. And learning the precise language of math comes with time, and with reading and writing math.

The vocabularies, grammars, and symbols—the languages—of STEM are critical to understanding the disciplines and becoming conversant in them. And as McSwiggen commented, becoming conversant in the field is the mark of understanding what it is that those who work in the field do. Across the STEM fields, tools are created and utilized, often for different purposes. The fields use different language and different types of texts. Literacy in the differing fields utilizes a variety of different tools for learning, yet there are similarities across them. By understanding the practices of the STEM disciplines, the importance of literacy within them becomes more visible, yet, should teachers and schools look for common strategies or disparate ones to teach all students? We believe there are different literacy tools that would work for each specific discipline, but there are also common literacy tools that should be taught in all STEM classrooms.

Challenges and Successes in Implementing Literacy Practices for STEM Disciplines

As we have mentioned throughout this chapter, there are numerous challenges that keep those in the fields of literacy and the STEM disciplines from providing the most beneficial learning opportunities to secondary students. The major disconnections or challenges, however, have produced opportunities that can be overcome if experts from both literacy and STEM disciplines work together to understand how they complement rather than conflict with one another. These challenges—breaking down disciplinary silos, changing teacher preparation, and establishing beneficial literacy practices for STEM classrooms in secondary schools—all have the potential for creating not only the right tools for the job, but also the right types of practices for utilizing these tools.

Breaking Down Silos

One challenge in particular is breaking down the disciplinary silos, which is the result of literacy experts not understanding what it means to be literate in particular STEM fields, as well STEM experts not understanding how literacy experts define literacy. Disciplinary experts often think of literacy in common terms, and that literacy is reading and writing, whereas many literacy experts recognize that to be literate, a person must join the "Discourse" (Gee, 2007) of the discipline. Thus, a challenge is to overcome this disparate understanding so that literacy and disciplinary experts can work together to facilitate learning within the STEM disciplines. In the STEM disciplines, connection to the real world, to other information, and to what is possible and probable is the defining characteristic. Current thinking for STEM initiatives often involves observation, inquiry, and discovery. These "tools" for the job, however, require that students understand what it means to embark upon inquiry that connects the disciplines in authentic ways; to understand the relationships between the STEM disciplines so discovery makes sense; and to observe phenomenon in a disciplined manner that addresses the questions being asked. To acquire this knowledge, however, requires students to know the major thinking of the disciplines address, the processes the disciplines undertake, and the language the disciplines speak. To gain that knowledge will take literacy experts and disciplinary experts working collaboratively to produce best literacy practices for particular disciplines.

From this study, we recognize the necessity of interdisciplinary collaborations between literacy educators and STEM experts. A major concept in literacy is that all teachers are "teachers of reading" (Gray, 1925). In essence, every teacher needs to be aware of the literacy needs within their disciplines. While literacy experts have felt that this was a requirement of all teachers, and this is reflected in many states' licensure requirement of at least one literacy course (usually content area or disciplinary literacy), it is seldom a requirement of literacy educators to be well-versed in disciplines other than language arts or English. While we are not suggesting that this become a requirement for literacy educators, it would serve the STEM and literacy fields to have a better understanding of literacy skills and strategies that are especially useful in the STEM disciplines.

One way this is happening is through co-teaching and team-teaching situations at the secondary and postsecondary levels. Another is through projects such as the T-STEM Network in Texas, which includes seven centers across the state that deliver professional development to teachers for transforming instruction in the STEM fields. In connection with this project there are also T-STEM Academies, secondary schools with a focus on STEM instruction. The Bill and Melinda Gates Foundation also is supporting the development of New Technology High Schools across the country, like the Academy of Technology, Engineering, Math, and Science (ATEMS) in Abilene, TX. Students at ATEMS engage in a PBL curriculum that integrates the STEM disciplines and requires them to develop the literacies necessary for work in the STEM fields.

As literacy experts collaborate with disciplinary experts in planning and implementing instruction, both learn from one another about their disparate fields and what literacy strategies would work best for facilitating learning. This practice is also happening at the teacher preparation, which will disrupt the current thinking of many teachers who assert they are only teachers of the discipline.

By breaking down silos and explicitly addressing what those in the STEM disciplines identify as specific knowledge and what would be considered literacy practices for their fields would create opportunities for teachers and teacher-preparation programs to move beyond generalist notions of content-area literacy. Through our study, we have found many content-literacy practices have little connection to the literacy strategies and skills necessary for understanding science, technology, engineering, and mathematics in authentic ways. We do not suggest, however, that the strategies or literacy tools developed and taught through a long tradition of content-area teacher preparation are without value. Instead, we suggest there is instruction that might be more relevant and useful to teachers as they prepare twenty-first-century students for careers in STEM, and teachers must become aware of those practices through the practices of all of those at universities who educate them to become teachers. This leads us to our second challenge in respect to disciplinary literacy and current practices.

Changing Teacher Preparation

A second challenge from this project applies to teacher preparation. Our disciplinary experts were convinced that the teaching and learning that is currently in K-12 content-area classrooms, along with the traditional textbooks for their disciplines, reflect little of the knowledge or the processes typically conducted in their fields. This is especially pervasive when addressing the integration of knowledge expected of STEM initiatives. Our experiences in middle and secondary classrooms confirm that teachers still rely upon the transmission of knowledge through lectures and demonstrations *about* the discipline, rather than the actual *doing* within the disciplines.

Our experts agreed that many of the university students who aspire to teach in the STEM fields view themselves as teachers *of* the field rather than content experts. Too often, secondary teachers lack confidence in their knowledge of the discipline. Their knowledge or lack of it thereof is typically not because they have not taken sufficient course work in the discipline, but the type of knowledge they learn is often not the metacognitive level. Thus, they have little knowledge of the major questions of the discipline, lack experience with the processes those in the field perform to construct new knowledge, or a firm grasp of the way language is used within the discipline. In essence, they are not familiar enough with the discourse of the STEM fields to consider themselves anything but novices in the disciplines. Thus, there is also a need to advise teacher candidates that the knowledge they learn in their content courses is necessary even if they feel they are *only* becoming teachers. Thus learning the discipline is important, as are the practices and thinking of that discipline. This challenge can only be overcome by disciplinary and literacy experts working together to establish greater expectations of teachers.

Our work suggests that to be literate in a discipline, teacher candidates must move beyond accumulating knowledge about the discipline and embrace the discipline's important theoretical ideas. They must come to a better understanding of the important questions and concepts of the discipline, and they will need to know how to seek answers to those questions. The best ways to acquire such knowledge—both demonstrative and procedural—is through research in the fields. Thus, teacher-preparation programs will need to build in undergraduate research as part of licensure requirements. Research within the fields would also address candidates' abilities to know what it means to read and write successfully within that discipline. Teachers of the STEM disciplines will need to be literate in these ways to apprentice students in grades 4–12. Frank Smith, one of the experts in literacy learning, suggests that young people need to become part of a "literacy club" (Smith, 1988), so they feel comfortable with literacy skills and strategies. The same is true of the STEM disciplines. Teachers need to apprentice their students in "STEM clubs" or clubs that address particular aspects of the STEM disciplines, so students become comfortable with STEM knowledge and processes. Thus, teacher-preparation programs need to promote, or even require, expertise of the literacy practices within the STEM disciplines.

Currently at the University of Cincinnati, there are multiple team and co-teaching situations at the postsecondary level through courses on content-area literacy as well as the methods courses for the STEM disciplines. As this work progresses, curricula is being developed with STEM colleagues and eventually will be based on the premise that students construct knowledge of a discipline by engaging in the literacy habits and practices valued and used by experts in the field.

Utilizing Beneficial Literacy Practices for the STEM Disciplines

While we have mentioned that generalist notions of content-area literacy are not beneficial for all STEM disciplines, one of the greatest challenges for literacy and STEM discipline experts is to discover which practices would be most beneficial for student learning within particular disciplines. Our project, along with others we have previously mentioned (e.g., O'Brien, Stewart, and Moje,1995; Draper et al., 2005; Moje, 2008), have addressed issues of curriculum and pedagogy in respect to the siloing that occurs within secondary schools. With each study, literacy experts are becoming more cognizant of the reality that they must work with disciplinary experts to increase their own expertise of disciplinary literacy practices. Even the change from "content-area literacy" to "disciplinary literacy" is a result of these studies and the changes within the field of literacy.

With our study, we found there are particular strategies that would be successful across all the STEM disciplines. Literacy instruction in

STEM-related classrooms would best serve the disciplines by aiming to build an understanding of how the field constructs knowledge, and thus, strategies such as thinking aloud, content-area literature circles (Johnson and Freedman, 2005), and guided observation are especially successful in building students' understanding of the discipline's discourse. Each of the following strategies are recognizable to our disciplinary experts as excellent tools for teacher educators and teacher candidates to use so their students will understand not only particular STEM disciplinary fields, but also are general tools used across the disciplines for greater opportunities for integrating STEM knowledge.

Note Taking—Knowing How to Look for Understanding

Throughout and across the STEM disciplines, observation plays a vital role in finding the answers to the questions of the field. Note taking is not a new way of learning in the STEM disciplines, but working *with* students through an interactive guided practice creates a foundation for their learning the content of the disciplines while also scaffolding the two other literacy strategies that we discuss in this chapter.

Students in secondary school, as well as teacher candidates, have difficulty knowing what they should be noting when observing a phenomenon or model. As our science experts mentioned, rarely do scientists make unfiltered observations but are often guided by a theory or pattern. They are looking for answers to particular questions. They do, however, record what does not fit the pattern or theory, which then has the potential for creating new knowledge.

In classrooms, teaching students to observe through a particular theory, to discuss how the phenomenon or model fits a particular pattern, or to address how the model or phenomenon doesn't fit a pattern or theory and why, would help students understand how STEM experts think about knowledge. Literacy strategies that would complement this type of observation would be interactive note taking, whereby the teacher and students work together to create appropriate notes when observing a phenomenon or situation. For instance, with this type of guided observation, in which the teacher combines thinking aloud while asking students to create observational notes to document a phenomenon, demonstrates how to observe a model through the lens of a particular theory. In addition, students have the opportunity—because they are being guided by the teacher to work through and then document does not fit a pattern. This then helps them learn what fits and does not fit, deepening their knowledge of both the pattern/theory and ways to observe. Case studies ask students to document a model or phenomenon by addressing essential elements such as conditions under which a phenomenon takes place or what elements need to be present for a model to be replicated. Students need to document how the model or phenomenon fits a pattern, what outliers were present, and what generalizations can be made from the case. With guided observation and interactive note taking of case studies, students can also learn how to examine examples and work toward generalizing knowledge that helps move the field—and their thinking—forward.

Literacy strategies that support this type of thinking also include advanced organizers, which scaffold the way to take notes. Teachers create the template for a case study or an observation that guides student thinking. An advanced organizer, created by the teacher, leads students' thinking by starting sentences that then allow the student to complete the necessary information. There is also structured note taking (Smith and Tompkins, 1988). Both of these types of scaffolding help students learn how to think about observing a phenomenon, creating habits of mind (Meier, 2002) that will serve them throughout school, and if they choose to enter a STEM career, throughout that career.

Thinking Aloud for Clear Understanding

Understanding the "what and how" of STEM knowledge is vital to joining "the club." Students can learn both when teachers demonstrate how they think about what is under study. Reading through a theory and explaining where it makes sense and why it makes sense facilitates students' learning. What may be more important, however, might be thinking aloud where some part of the theory can be misunderstood or misinterpreted and demonstrating the question-posing that comes with attempting to remedy a misconception. Once students become familiar with the thinking-aloud strategy, they can practice this same process in small groups or pairs where students can work together to address misconceptions or questions of understanding. Through a strategy called content-area literature circles, students can discuss their thinking, their misconceptions, their mathematical processing, and their "what if?" type of questions that often accompany STEM thinking.

Content-Area Literature Circles—Talking through Our Understanding

Content-area literature circles are small group discussions that allow students to discuss the information they have read or the notes they have taken during a phenomenon. Content-area literature circles are typically guided by particular prompts that make sense to the discipline, and thus, teachers have the ability to create habits of mind that address the major understandings of the field, the processes students should learn in respect to the field, the simulations they created or observed, and use the language that is vital to understanding and being literate in the field. Other times, however, content-area literature circles can behave like literature circles in language arts, which allow students to discuss their burgeoning understandings of the content. With such discussions, misconceptions can be addressed, "what if?" thinking can be practiced, and deeper understandings can be developed.

Occasionally, teachers feel uncomfortable with student discussion in small groups. Yet, according to all the STEM experts, being able to discuss their content and concepts is one of the most important attributes of being literate in their fields. Participating in types of talk common in their particular fields, along with keeping current in the field through reading, are necessary skills for gaining a deeper knowledge of the STEM fields. Content-area literature circles can facilitate both of these skills. Generally, teachers can accommodate content-area literature circles in their classrooms by allowing for short periods of time and with particular prompts that focus on the daily objectives. Teachers can circulate throughout the room and take notes on what students discuss, the misconceptions that may be part of student thinking, or the lack of understanding of particular models or theories. From there, additional teaching may be necessary, but then teachers are assured of their students' learning.

What is essential to note in respect to these strategies is that they are complementary, and in many ways build upon one another. Asking students to take notes through guided observations and then discuss them in content-area literature circles using a think-aloud strategy allows them to not only address what they are learning, but also to discuss it with peers in ways that help them see where their thinking is on track and where there could be misconceptions. In addition, to work through "what if?" questions, which are vital parts of STEM initiatives, students come to realize that the STEM disciplines are not about learning facts, although foundational knowledge is essential in being literate in the field. Students learn that the STEM disciplines are about discovery and progress.

We have come to realize the value of exploring the topic of disciplinary literacy and the potential these explorations hold for improving the opportunities adolescents may have for constructing knowledge of content and practice in the content areas. As literacy experts, we were surprised by the common literacy strategies we found where we least expected them. Our math colleagues talked extensively of the value of asking students to talk through their processes for problem solving, an activity immediately recognizable to us as a think-aloud whereby students would work in small groups or circles and talk through their mathematical processes. With science experts, the need for guiding students' observations was important. And, with technology, our colleagues stressed the need for students to work collaboratively in locating, organizing, and using information from a variety of sources in search of solutions to open-ended problems. This framework corresponds closely to the inquiry/research strategies taught in literacy curriculums.

At the STEM level, we found that language is vitally important, regardless of what discipline. We also need to build a common vocabulary to discuss our interdisciplinary understandings. This is not to disrupt the vocabulary of the discipline, but a practice—a new discourse—that allows us to talk a common language that honors our disciplinary differences while also building on our connections. Our STEM colleagues recognized and welcomed the literacy strategies we described as appropriate for students as they learn to *do* the disciplines. Through this project we began to see that disciplinary literacy, rather than being irrelevant, can help move teachers closer to *doing* rather than reporting about the discipline they teach.

Conclusion

For years, content-literacy educators have been guilty of suggesting strategies and texts for content learning, but have done so with little or no consultation with disciplinary experts. With STEM content—as individual disciplines or as integrated projects—that continues to develop and produce new knowledge, literacy initiatives must cultivate a new relationship with these disciplines so as to create the best learning strategies available. Through our work with colleagues in the STEM disciplines, we have the opportunity to produce results that will not only increase student learning of STEM knowledge, but also to create students who will have the capacity for STEM leadership in the future.

We have only started to explore the ways in which literacy and the STEM disciplines can work together. Our next steps will include developing further strategies and skills available for student learning. In addition, these steps will involve ways in which the STEM disciplines are connected to a greater number of knowledge networks to understand and implement expanded ways of knowing what is possible for extending our sustainable efforts for generations to come. These steps will broaden what it means to know, to be literate, and to further develop the current knowledge of a greater breadth of disciplines. This expansion is a crucial step in creating an infrastructure that will allow us to work across disciplines to better educate our citizenry. We also need to conduct further research with teachers as they implement particular disciplinary strategies in their classrooms. We should continue building relationships that bridge the gaps between content knowledge and disciplinary literacy with the ultimate aim of understanding that to be literate in a discipline we must construct content knowledge through the practices and habits of mind common to that field.

Chapter 6

A Case Study of a STEM Grant's Executive Board: Challenges with Ownership and Initiative

Abdulkadir Demir, Camille Sutton-Brown, Lacey Strickler, and Charlene M. Czerniak

The transition to a globalized world has led us to the realization that Science, Technology, Engineering, and Mathematics (STEM) subject areas are critical for prosperity in a knowledge-based economy (Committee on Science, Engineering and Public Policy [CSEPP], 2006; Matthews, 2007). The United States needs to fill positions in science and engineering with individuals who are scientifically literate. Local and national job projections all point to growth in careers in sectors such as healthcare, information technology, engineering, and manufacturing that require a background in science and mathematics (Ohio Department of Job and Family Services, 2006; Bureau of Labor Statistics, 2008).

Although the United States is currently competitive in new scientific discoveries, the potential to continue at the forefront of discovery is not certain. The CSEPP published *Rising Above the Gathering Storm: Energizing and Employing America for a Brighter Economic Future,* and in this publication the committee discussed their concern that the "scientific and technological building blocks critical to our economic leadership are eroding at a time when many other nations are gathering strength" (CSEPP, 2006, p. 3). Similarly, the National Science Board (NSB) recently released a report entitled *Science and Engineering Indicators 2010* that shows the United States continues to be a leader in the knowledge and technology intensive (KTI) industries at this point in time, but its position in the global economy in these industries has begun to flatten and slip, and the future for the United States in the KTI industries is uncertain (NSB, 2010).

In response to these types of reports, public decision makers have pushed to make science and mathematics a top educational priority in the United States. Paralleling the suggestions put forth in *Rising Above the Gathering Storm* (CSEPP, 2006), national leaders in education and government have developed priorities for STEM education (Matthews, 2007). Over the last decade, these national priorities have evolved from several influential policy reports demanding comprehensive changes in science teaching and learning. Several of these reports include Project 2061 developed by the American Association for the Advancement of Science (AAAS) (AAAS, 1989), the *National Science Education Standards* (NSES) developed by the National Research Council (NRC) (NRC, 1996), and America 2000 (U. S. Department of Education, 1991) developed by a committee of the nation's governors. Together, the recommendations aim to prepare a scientifically literate national workforce that is prepared to compete in an increasingly scientifically and technologically oriented global economy.

The current focus on STEM in the United States is not new. Science and mathematics education reform is a topic that has been discussed since the days of the "*Sputnik* Challenge." Why have reform efforts not been successful at reinventing science and math education? This question is difficult to answer. In order to gain a better understanding of why reform efforts have been less than successful, we believe it is important to discuss the literature related to STEM reform initiatives, K-12 school reforms, and organizational change.

Literature Review

The United States is currently in the midst of a large-scale reform effort in STEM education that requires changes in curriculum, teaching practices, and community involvement. Here we address various STEM reform initiatives, K-12 school reforms, and organizational change.

STEM Reform Initiatives

The Carnegie Foundation report entitled *Opportunity Equation* (Carnegie, 2009) recommends focusing on four priority areas: (a) higher levels of mathematics and science learning for all American students; (b) common standards in mathematics and science that are fewer, clearer, and higher coupled with aligned assessments; (c) improved teaching and professional learning, supported by better school and system management; and (d) new designs for schools and systems to deliver mathematics and science learning more effectively.

The Obama-Biden U.S. presidential administration proposed a plan that prioritizes mathematics and science instruction in the attempt to prepare young citizens to be active members of a technologically dependent society (Obama for America, 2009). In January of 2009, the NSB, the national science and engineering policy advisory body transition team for President Obama, released it's STEM education recommendations that included the need for motivating the public, students and parents, clear educational goals and assessments, high-quality teachers, world-class resources and assistance for teachers, early STEM education, local excellence, and national coherence in STEM education (NSB, 2009).

STEM education reform is also at the epicenter of several coalitions, councils, and university initiatives. Most notably, the Triangle Coalition has been in existence for over a decade, and it focuses on bringing together the voices of government, business, and education to improve STEM education. More recently, the STEM Education Coalition has aimed to strengthen STEM-related programs and increase federal investments in STEM education. Science and Mathematics Teacher Imperative (SMTI) consist of public university leaders that aim to catalyze action for STEM education across state and federal governments, businesses, and the K-12 community. The membership of these organizations consists of teachers, researchers, scientists, business leaders, universities, and professional organizations. The groups have in common the objective of bringing together business, education, and government to improve STEM education and to support new and innovative initiatives and STEM-education policy.

At the federal level, funding for STEM education has increased considerably. In November of 2009, the White House released the launch of the "Educate to Innovate" campaign meant to create partnerships between companies, foundations, nonprofits, and science and engineering societies with the goal of motivating students across America to excel in science and math (The White House, 2009). Recently, President Obama expanded the campaign to include a new goal of attracting, developing, rewarding, and retaining teachers in science and math education (The White House, 2010).

Federal funding for STEM education programs at the U.S. Department of Education increased from 12.5 million in 2002 to 182 million in 2007 (Triangle Coalition, 2005; U.S. Department of Education, 2010). As one example, the

National Science Foundation's Mathematics and Science Partnership (MSP) grants provide funding to a variety of teacher leadership, statewide, and targeted STEM programs that aim to advance academic achievement in mathematics and science. Funding for the MSP programs was 60.99 million in the year 2009 (NSF, 2010).

One statewide MSP Georgia, *Partnership for Reform in Science and Mathematics (PRISM),* is focusing on a preK-16 approach to education and seeks to increase science and mathematics curriculum, raising public awareness for science and mathematics education, increasing and sustaining qualified teachers and increasing higher educations role in the schools, and influencing statewide policy (Kutal et al., 2009). *PRISM* is currently in phase II in which they are conducting research to determine the strategies that lead to changes in policies, public awareness, practices, partnerships, and resources for science and mathematics education (Kutal et al., 2009).

K-12 School Reform

Reforms in STEM at the K-12 level are under way across the nation. Recently, most of the 50 states scrambled for *Race to the Top* funding for school improvements, and 12 states have been awarded funding. *Race to the Top* is a competitive grant program designed to encourage and reward states that are working toward improving graduation rates, better preparing students for college, closing achievement gaps, and improving student outcomes (U.S. Department of Education, 2009).

How are schools approaching these various reform efforts? There are several ways to discuss school reform and several suggestions as to how to implement it. Some of the most common theories to accomplish school reform focus on standards-based district-wide reform initiatives, professional development of teachers and administrators, and retention of quality leaders both in administration and teaching (Rutherford, 2005). Indeed, these were the focus of the No Child Left Behind Act (U.S. Department of Education, 2002) and are the focus of the new *Race to the Top* efforts under way now. Changes in curriculum, teacher preparation, teacher professional development, and increased participation of community members in the schools are just a few of the suggestions for implementing and sustaining school reform (Committee for Economic Development, 2003).

One model for school reform is the Dynamic Model of Organizational Support presented in 2003 by Gamoran and Anderson. The Dynamic Model of Organizational Support focuses on teacher professional development to improve student understanding of science and mathematics. This model concentrates on what is happening inside the school and the dynamics between teachers, administrators, and students. This model attempts to analyze how districts and schools support and challenge teacher professional development in order to improve student understanding.

Role of the Community in School Reform

Fullan (2006) contends that change theories focusing on curriculum and teachers are useful but incomplete in that they do not get close to what happens in classrooms and school cultures, and that they focus too much on producing better individuals as a way to change the system. It has been argued that in order to sustain change, schools and communities must work together and move to an era of collective collaboration and collective accountability for science and mathematics education (Zmuda, Kuklis, and Kline, 2004). Key stakeholders such as teachers, administrators, parents, and community members need to be actively involved and understand the direction of the school reform (Gamoran et al., 2003; Zmuda, Kuklis, and Kline, 2004).

The NRC supports the notion that change cannot occur by looking only at schools and stresses the importance of community members outside the immediate education system such as scientists, businesspeople, parents, engineers, legislators, and other public officials to achieve the implementation of the *NSES* (NRC, 1996). The *Project 2061 Blueprints for Reform* also discusses the importance of a highly integrated network of parents, community members, parents, and educators for the success of children in science and mathematics (American Association for the Advancement of Science [AAAS], 1997).

Fullan (2006) discusses a theory of action for school reform that consists of seven premises, which are: a focus on motivation; capacity building, with a focus on results; learning in context; changing context; a bias for reflective action; tri-level engagement; and persistence and flexibility in staying the course. What is interesting here is the sixth premise of tri-level engagement, which refers to school and community, district, and state.

Community organizing can be used to catalyze school reform and has been the subject of increased attention within the last ten years (Shirley, 2009). Community organizing groups for school reform generally exhibit specific characteristics that distinguish them from legal-aid groups, advocacy groups, social services, et cetera. These characteristics include:

> They work to change public schools to make them more equitable and effective for all students; they build a large base of members who take collective

> action to further their agenda; build relationships and collective responsibility by identifying shared concerns and create alliances and coalitions to cross institutional boundaries; develop leadership among community residents to carry out agendas and use strategies of adult education, civic participation, public action and negotiation to build power for residents of the community that results in action to address their concerns (Gold, Simon, and Brown, 2002, p. 14).

One example of community involvement in school reform is a coalition of over one hundred schools named Alliance Schools. One community organizing strategy they used was the use of accountability sessions with business and community leaders. These accountability sessions enlisted the help of business and community leaders to engage parents with the school to improve parent involvement (Shirley, 1997). *The Cross City Campaign for Urban School Reform* also documented several aspects of community organizing and has developed an Education Organizing Indicators Framework that educators, organizers, and funders can use to understand the impact of community organizing on school reform (Gold, Simon, and Brown, 2002). An important piece of that framework is that community organizing sought to broaden out the issue of accountability by taking the burden of accountability for student academic success away from just the teachers, students, and parents and broadens it out to an array of other stakeholders. This public accountability is what connects school improvement with the community as a whole, and it creates the will to continue the school/community connection that can ultimately play a role in school reform (Gold, Simon, and Brown,, 2002).

Organizational Change

The topic of school reform and change ultimately leads to a discussion of organizational change. Schools, as organizations, must continually reevaluate what is important and restructure accordingly. School reform may not be successful or sustainable due to idea that the organizational change that is required is more radical than the changes that are taking place (Fullen, 1995).

Organizations that are stable will exhibit lower failure rates and will be preferred over organizations that are able to change readily (Hannan and Freeman, 1984). However, when organizations find themselves in a rapidly changing environment, adjustments in organizational change prove beneficial to their short- and long-term success (Haveman, 1992). So, organizations will have resistance to change due to the need to be reliable and dependable to produce or provide their particular services, however, if there is a significant change in the environment in which the organization resides then it is beneficial that the organization has the ability to make changes to its structure.

Mai and Akerson (2003) look to leadership roles as a route to organizational change, specifically the roles of *critic/provocateur* and of *learning advocate / innovation coach*. The role of *critic* and *provocateur* is one in which the leader makes it safe to question the practices that an organization uses to reach its goals. This role helps to create a safe environment for all members of the organization to reevaluate the status quo. The role of the *learning advocate / innovation coach* is one in which the leader actively supports learning that allows for team problem solving, use of data to guide practices, and the sharing of knowledge. This role of the leader relies on two strategies, facilitating discussion and encouraging innovation (Mai, 2004). "These roles and the behavioral attributes describing them warrant consideration by both school administrators and teachers who want to address the challenge of continuously improving educational practice in their schools" (Mai, 2004, p. 221).

In summary, school reform needs to happen from the inside and outside. The schools are organizations that have been reliable and stable but current environmental conditions require the schools to change in order to be successful. This change needs to be implemented through leadership from administration, teachers, students, parents, districts, and community involvement/organizing. Community leaders can serve as *provocateurs* and *advocates* for STEM education reform, and it is with this in mind that the grant-funded program described in this book included the formation of an Executive Board consisting of key community leaders.

Context of the Study

This research was completed at a Midwestern public Doctoral/Research Extensive University that received a U.S. Department of Education Teacher Quality Enhancement (TQE) Partnership grant[1] in 2004, which was designed to recruit, better prepare, and retain science and mathematics teachers for urban schools. The grant-funded program included many activities (e.g., scholarships, paid internships, new courses for preservice teacher candidates, new courses for mentor teachers, and seminars for principals). One component of the program focused on public-relations strategies aimed at making the community aware of the importance of STEM education and the need for highly qualified science and mathematics teachers.

An Executive Board including prominent community leaders in education, business, and local and state government (e.g., superintendent, school district union president, state senator, deans of education and arts and sciences, business leaders, director of the zoo, director of the local science museum) was formed to provide oversight to the grant and to help secure funding when the program ended. However, soon after the Executive Board was formed, the members strayed quickly from the charge as described in the grant. Most members stated that they felt the Principal Investigator (PI) was more than capable of running the grant, so they did not feel the need to oversee the grant's progress. Further, they felt that fund raising was the prerogative of the University Foundation Office. As the members were key influential members of the community, they were more interested in community awareness activities and community advocacy for STEM education. The Executive Board was interested in serving in two roles: to serve as a catalyst for change in STEM education in the community and to advocate for STEM.

In other words, they ended up serving as *provocateurs* and *advocates* for STEM education reform. The Executive Board met once per semester and engaged in a number of activities including writing editorial letters about the importance of STEM to the community in the local newspaper, hosting a community STEM summit that included over hundred attendees from K-12 education, university, parents, business, and community leaders, and forming community work groups that focused on specific aspects of STEM education (e.g., K-12, higher education, community engagement).

Purpose of the Study

The primary goal of the Executive Board as the focus of this study was to change public, political, and private support and awareness in the community regarding the importance of STEM teacher education and STEM achievement in K-12 schools. A closer look at the functions and responsibilities of the Executive Board derived primarily from this goal become an important case to examine. This study was designed to answer the research questions listed below.

Overarching Question

What factors are associated with a grant-funded Executive Board's ability to create the private, public, and political will to improve science and mathematics education and impact policy in the community?

Sub-Questions

1. What are the key issues of K-12 STEM talent development and STEM teacher education in the community that concern Executive Board members? How do these issues and concerns influence Executive Board member's decisions to get involved in the TQE initiative?
2. How can the Executive Board serve as a catalyst and advocate for strategies that sustain an ongoing system to improve the quality of K-12 science and mathematics education in the community?
3. What key groups of people can be counted on to assist in the program initiative to convey the importance of STEM fields to a number of different constituents including private, public, and political audiences? Or making science and mathematics teaching more attractive?
4. What are the constraints and challenges of developing a functioning Executive Board aiming to initiate and implement improvements K-12 STEM talent development and STEM teacher education in community?

Research Design

The research design framing this study used a qualitative case study approach (Yin, 2003) that emphasized participants' roles as board members, as well as their experiences and perceptions of the grant-funded Executive Board. The case study was phenomenographical in that it focused on developing, recognizing, describing, and apprehending the qualitatively different ways in which people experience certain phenomena or certain aspects of the world around them (Marton, 1981, 1992). The focus of phenomenographic research is to find the variation that differentiates the phenomenon for participants, rather than finding the singular essence (Marton, 1996). The phenomenon under study was the functionality of the TQE grant's Executive Board. In this study, *case* refers to the total group of board member participants ($n = 13$). The phenomenographic approach applied to case study is grounded in the interpretative research tradition and provides insight into the functions of an Executive Board in the context of federally funded TQE grant initiative.

Participants

The 26 members of the TQE grant Executive Board were residents of the urban community, represented a myriad of academic and professional disciplines, and had varied backgrounds. Representation included a physician, a psychologist, college deans, a former university president, a state senator, former K-12 teachers, K-12 school administrators, university personnel, school board and union members, as well as community members, such as church members and parents. There was also representation from the business sector, including CEOs of local companies and public affairs directors. This illustrates that this board was comprised of various people in the community who hold enough political and economic power to advocate for adjustments in the education system. The wide range of diversity among the members allowed the board to draw upon varying experiences and perspectives in its attempts to design and implement various initiatives. All of the members had an interest in STEM issues, which was why they were members of the board. The participants of this study included the TQE grant's Executive Board's director and 12 members who were purposefully selected based on their active participation on the Executive Board (see table 6.1).

Table 6.1 Participants and Their Role(s) in the Community

Participant Name	Role in the Community
Dr. Thompson	Associate Dean for natural sciences and mathematics, interim dean for graduate studies
Dr. Belmont	Executive director of a zoo
Dr. Lange	Dean of College of Arts and Science
Dr. Young	Dean of College of Education
Dr. Washington	Physician pathologist—the chair to the board at a local Catholic boys high school and the chair of the board at the United Way
Dr. Carter	Director of marketing at a local community college, PhD in education
Wilson, Shawn	President/CEO of the a local economic development organization
Dr. Matthews	Dr. Matthews is the vice president for government relations at a local university and is also the chief of staff in the president's office at that university
Malone, Derrick	President of a local association for administrative K-12 school personnel
Simpson, Opal	Public affairs director of a refinery, member of a local University's board of trustees
Fredericks, Tracey	Local state senator
Rodriguez, Richard	President of an engineering company
Dr. Chamberlain	Education professor at the university

Data-Collection Procedures

All primary data for this study were collected using a semi-structured, open-ended, interview protocol (Seidman, 1998). The purpose of this type of interview is to have research participants reflect on their experiences, and then relate those experiences to the interviewer in such a way that the two come to a reciprocal understanding about the meanings of the experiences (Marton, 1994). The primary researcher (hereafter referred as *the researcher*) completed one audio-taped interview with each research participant to gain access to their experiences with Executive Board as a member and their understanding of the board's primary goal. During the interviews, the researcher focused on the research participants' background and vision and perspective of the Executive Board, along with perceptions of STEM issues in the urban community. The one exception to this is the interview with the TQE grant director/PI, who was also asked questions regarding the current functioning of the Executive Board, her expectations of the board members, along with the challenges to effectiveness of the board. Each interview lasted approximately 45 to 75 minutes. All interviews were transcribed verbatim and in full and reviewed twice for accuracy before analysis. Interview questions are provided in appendix 6.1. To gain full knowledge of the participants' experiences, we also utilized data from different sources. These additional sources were essential to examine any discrepancies between participants' statements and their actions if there were any. These included written documents, such as meeting minutes and emails.

Data Analysis

We analyzed the transcripts as much as possible from a tabula rasa, seeking what emerged as important and of interest from the text. We started our data analysis by examining the combined interview data of 12 board members. Following this, we analyzed the data from the interview with the director, considering her data individually and the data from the other participants as a collective. For all of the transcripts, the central focus of the data analysis was on differentiating parts of the data "in terms of their internal consistency (do the elements show consistency in the referential and structural aspect), and the relations between them (do they together provide full coverage of variations in the total data)" (Hasselgren and Beach, 1997, p. 194). With the purpose of determining the categories of description and their internal consistency, we applied an inductive data-analysis process, grounded in the utterances of the

Executive Board members and iterative procedures (Patton, 2002). We read and analyzed the transcripts to determine a set of analysis codes and then developed a set of claims. In order to confirm the claims and "to reduce the likelihood of misinterpretation" (Denzin and Lincoln, 2005, p. 453), we utilized a number of illustrative examples. To increase the credibility of our findings during the analysis phase, we used triangulation of data sources and analysis of disconfirming cases.

The research questions that framed this study guided us as we examined the primary-level codes. Two researchers from the research team organized the data and created categories of descriptions across the data set. These two researchers also took the lead in coding data, bringing the codes and data from which the codes were constructed to the other researchers for examination and revision. As data analysis progressed, the researchers shared their developing understandings of the data. Ultimately, the number of categories was reduced by merging and eliminating categories and by clustering others based on recognized connections (Attinasi, 1991). The researchers resolved any disagreements in category assignments through further discussion. Data codes included categories such as board members' perceptions of STEM education and structure and function of the Executive Board, suggestions for improving the current functioning of the Executive Board, among others. The following section is a summary of the major findings of the study.

Findings

This findings section is organized according to the main themes that emerged from the data. The findings are presented in four sections. First, we discuss the board members' perceptions of STEM issues. In the second section, we present the findings relevant to the structure and functions of the Executive Board. The third section presents the director's perspective of the board and STEM issues. The fourth section includes the findings from the meeting minutes and the accomplishments of the Executive Board to date.

Perceptions of STEM Education

This section first presents the Executive Board member's perceptions of STEM education. Specifically, we present their stated beliefs about the importance of STEM education in general, the impediments to STEM education, and their proposed solutions to the identified impediments. Each of these topics is discussed in depth.

Importance of STEM Education

Every board member, regardless of their affiliation, acknowledged that STEM education is very important. They based this importance on the crucial role of STEM fields in today's society, specifically in reference to the global competitiveness in that exists in the sciences and technology. An example of this claim is reflected in Mr. Rodriguez's statement that

> [STEM education] matters because, in a world, a global world, where technology is quickly outdated and obsolete, its individuals developing new, innovative approaches at many different levels, that create the economic activity necessary to grow both a national, but an international environment that continues to provide value. And the basics begin in the classroom, through the education of our youth, in science and mathematics. If that does not happen, then you are not going to have the innovation, nor the economic growth that you need in a society.

Additionally, Dr. Washington points out that STEM knowledge is necessary for all fields, not just the obvious ones such as engineering or scientific research, for it permeates all forms of industry in the United States, health care, medicine, business, and the service industry. Therefore, the board members assert that to remain a top contender in the global market, American workers must be highly trained in this field in order for the United States to remain one of the key global players in this arena. Also, as computers and other forms of technology are increasingly becoming more integrated in society, thus to function independently and to be a contributing member to society, individuals will need to have at least a basic level of scientific literacy.

Impediments to STEM Education

Recognizing the need for citizens who have a working knowledge of STEM skills, the board members express concern about the lack of talent at the K-12 as well as the university level. Many of them noted that implementing a STEM initiative at the university level is not the most effective approach, as the problem originates earlier, at the primary level. Dr. Belmont, the executive director of the local zoo, asserts that

> if you have someone who is graduating from high school and about to start college, and they have never been interested in science or math, you are not

> going to make them choose a math or science career. Because they have no interest! So I think it has to start way before that. I think it starts back in 3rd, 4th, 5th grade where you keep kids interested in STEM education.

A few of the reasons that they cited in reference to this problem was attributed to the negative societal perceptions of science and mathematics. They felt that society associated math and science with difficulty, and this reflects in the lack of student motivation to learn scientific and math skills coupled with lack of parental involvement that they have witnessed. Another key contributing factor was the limited amount of school resources to provide the students with opportunities to apply the skills in a hands-on manner.

Of major concern was the quality of teacher education and preparation to teach science and math. The majority of the board members stated that teachers do not have an adequate amount of content knowledge to teach STEM subjects to students. Due to their limited understanding of the subjects, they do not expose students to the full range of issues, and they cannot present the information well, because they are not as comfortable with the subject matter compared to the humanities subjects. Those who graduate with university degrees in STEM subjects often go into other career fields, because the salaries offered to teachers are low and not as attractive as other business ventures. This leaves vacancies for math and science teachers, which are often filled by teachers who have been trained in other disciplines, and who do not have a solid background knowledge of STEM subjects. As a result of teachers' limited content and pedagogical knowledge and preparation, K-12 schools do not focus enough on the STEM subject areas.

> There is an undue, maybe, emphasis on methods of teaching, but not on the subject that the teachers teach. I have seen it first hand with teachers in high school or even elementary school when they teach a subject that they are not comfortable with and, if they are not comfortable, a student who listens to them senses the discomfort and it is something that is transmitted to the students that this is a subject that is not a comfortable subject, so we have to make sure that the teachers have not just sufficient, but more than sufficient understanding of the subject matter that they teach. I have seen people teaching math and/or physics with a major in geography. (Dr. Thompson).

The board members suggested that there is a relationship between teacher pedagogy and student interest in science and math. Dr. Washington, for example, noted that teachers who have degrees in education do not necessarily have a strong content background in math and science, due to the limited requirements. This impacts their ability to effectively teach these subjects to their students, resulting in students' low motivation to acquire these skills. The unfortunate outcome is that the students do not master the fundamental math skills necessary to advance to higher level math and science courses at the university level and avoid taking math and science beyond the minimum requirements in high school.

Proposed Solutions to Identified Impediments

To address the above stated concerns, the board members offered ideas for improvement. Increased salaries, scholarships for additional training in content areas, and other incentives were offered as ways to address the issue of quality of teacher content knowledge and preparation to teach STEM subjects, along with higher admission criteria for teacher education programs. This would help to change society's perception of educators, as teaching is no longer a highly regarded profession. It would also promote better streamlining between teacher training and practice, particularly with helping teachers learn how to incorporate math and science with liberal arts and humanities subjects in an interdisciplinary manner. They suggested finding innovate ways to sustain parental and student interest in STEM subjects, starting as early as kindergarten. Increased funding for school resources through community and business partnerships, reinforcing STEM skills within arts and humanity courses, and creating specialized schools that focus primarily on STEM subjects were other suggested ways to address this issue.

Senator Fredericks, a former elementary school teacher, proposed a model that summarizes key strategies to improve the quality of STEM education throughout K-16. She emphasized the crucial role that teachers play in providing students with foundational knowledge and mathematics and science. "Providing quality teachers is key and good teachers matter," she says before stating the importance of federal and state government–funded incentives for individuals with math and science backgrounds to pursue a career in teaching. However, education should be approached holistically; therefore, we must extend beyond the teacher's role to effectively improve the quality of education. She stressed the need for community and parental involvement in public education, stating that "we need to have a strong involvement with mentors in our community—reaching out to the school systems, reaching out to parents, reaching out to children, and, in the fields of Math and Science." She believed that if the parents are educated about the available jobs currently in the market, and the high demand for STEM skills, that they will be instrumental in motivating students to excel in these subject areas. Student motivation is also affected by their opportunities to apply their learned skills, therefore, she recommends that internships at the high school level are created for students to engage in real-life situations where they can apply their knowledge and connect them to the jobs of the future.

> Children are bored. I think because they have learned division for eight years. If you have learned to divide, why do you have to have eight years of learning how to divide? You need to incorporate those basic skills in math and in science and put them into real life experiences. Have children involved in exploring why they are learning math, why are they learning science, and how does this translate into the real world. This will make it more meaningful for them and for the teachers.

Therefore, teachers must also be exposed to real-world applications of the science and math that they are teaching. This is should be an important component of their preparation programs, but it should not end there. Professional development should also include this kind of engagement. Senator Fredericks proposes a collaboration between different parties, including teachers, businesses, and the larger community. "I think that [this engagement] could happen between the industry that focuses on Math and Science and build partnerships within the community so that the teachers see how that Math and Science is relevant to the technology that they are teaching, because I think that there's a disconnect between educators and the real work world," she commented.

None of the propositions can happen in isolation, however. The crux of Fredericks's model can be realized only with support from federal and state governments. Financial resources are needed for teacher incentives, improved labs in schools, and the provision of internships. Thus, teachers, parents, and students are key players; however, the broader community must also be actively involved. One such example is the TQE grant Executive Board.

Perceptions of the Structure and Function of the TQE Grant Executive Board

Since STEM teacher education is perceived as being important, the Executive Board was initially formed to oversee the grant and conduct fund raising for sustainability of the grant activities, as previously mentioned. As such, the director strategically sought specific people for a position on the board. However, the members preferred to serve as STEM advocates. The structure of the board is discussed specifically in reference to the nature of the members' involvement in the board. We then turn to a discussion of the function of the board, and this is presented according to the three main subsections. We first discuss the perceived goals of the

Executive Board, followed by a discussion of their perceptions of how an ideal Executive Board would be run. Lastly, we present their suggestions for improving the current functioning of the Executive Board.

The Members Involvement in the Executive Board

The director selectively chose specific people to serve on the board based on their status in the university and the community. Most of the board members stated that they agreed to participate in the Executive Board out of the respect that they have for the director, herself very well known and accomplished in the community and nationally and internationally. In addition, despite their career field, each of them expressed having a personal interest and commitment to improving STEM education in their community. Dr. Matthews illustrates this in the statement that "besides Cynthia asking me, it's my interest and desire to continue to promote education like I have done all of my life." From a business perspective, Steve Wilson, a CEO of regional economic development, explained that it was a natural fit for him to be involved in a STEM-related initiative, as it supports the growth of local science and technology businesses. "If we are trying to attract or grow a technology company, they are going to be asking 'where are the scientists, where are the engineers, where are the mathematicians?'" Therefore, in his opinion, it is important to support STEM initiatives to attract and retain businesses to the area that will contribute to economic growth.

At the individual level, each interviewee expressed a personal agenda that they had. They viewed the board as an appropriate avenue to advocate for their own concerns and to accomplish the various goals that they had set. For example, Dr. Belmont, the executive director of the zoo, wanted to promote the services that the zoo offers. By participating in the Executive Board, she could garner publicity for the educational programs that are offered at the zoo and other local attractions. Thus, she seeks to solicit more teachers to utilize these resources in the context of experiential learning via field trips. She says,

> The science center, the zoo, the metro parks, and botanical gardens are places that you can teach kids outside of classrooms. They are places that kids really can get involved in experiential learning, and I think we need to help teachers do a better job of using those resources.

Perceived Goals of the Executive Board

Although they all joined the board for personal and professional reasons, the interviewed members were unclear about the Executive Board's goals and purpose. There was not an agreed-upon purpose of the Executive Board, and two members admitted that they did not know its "real" purpose. "It is sort of a problem for me to answer that, because I am not sure yet, what the exact purpose is. I do not know the goals, exactly," Dr. Matthews admits. Some descriptions that the other board members gave are that the board is a group for strategic planning, it is a support mechanism for the overall TQE grant initiative, it is a means to promote community awareness about STEM education, and being more proactive in terms of publicity surrounding STEM issues in Metro Toledo Area, and that it is a means for fulfilling the requirement for the TQE grant by doing what was proposed in the grant application. This lack of uniformity among respondents suggests that the purpose of the board was not explicitly communicated to the members.

Board Members Perceptions of An Ideal Executive Board

All of the interviewed board members cited previous experience with other boards, and they draw upon those experiences to assess the current functioning of the TQE grant Executive Board. They each had their own perceptions of what a functioning Executive Board entails, but the commonality among their responses is that the TQE board is not currently an effectively functioning board. They cited a few reasons for this, including that the goals of the board are not clear to the members, the board does not meet often enough, and that it has not accomplished much to date. Despite the current status of the board, the members remain hopeful in what the board will accomplish, if changes are made to its systematic functioning. They stated the intended accomplishments according to time frame. The long-term goals is to help to actively promote STEM fields in K-16 in an attempt to increase the number of graduates in these fields. The short-term goals are to support programs that help to improve the quality of teacher training in the math and sciences as well as to increase the amount of resources available to K-12 schools for these subjects.

Board Members Suggestions for Improving the Current Functioning of the Executive Board

The board members, when asked about how the functioning of the board could be improved, offered some suggestions. They believed that the number one priority should be communicating the purpose and intended objectives to the board members. Related to this, each board member should be made aware of, or take initiative to determine, their specific role on the board. Dr. Lange, a dean of the college of arts and sciences, recommends that the Executive Board should advise the direction and strategic planning of the TQE grant in a way that it provides advisement for this entity. He elaborates on this point by saying that the board must "understand the director's direction and expectations as well as the sense of the board's mission. If the mission and goals are near to us, then the board can help to implement or help to apply and plan the direction."

Another means of improvement is to establish a more participatory model of decision making. Currently, the leadership style of the board is less authoritative than the interviewed members prefer, and that they would like more input into planning the activities that the board engages in. Malone compares the board meetings to formal sessions in school. He explains that people only speak and provide input when they are asked, as it is not an open forum for each member to express their ideas. Dr. Belmont echoes the sentiments of the other board members in that when they do have opportunities for input, the members are unclear as to their role and what they should be contributing, because the leader does not clearly articulate demands. Simpson states, "I do not feel involved or engaged in [the board] enough to be of help," paralleling Dr. Washington's comment that "I have no idea what my duties are. The only thing I know is that I get an occasional notice to come to a meeting." He advises that, "there has to be more structure. There has to be more specific requests of board members", such as "who is going to be assigned what task to make sure you get from point A to B to C to D." If not, he cautions, the board's efforts will be in vain.

The majority of the interviewed board members stated that they do not meet often enough. This inhibits their ability to know each other and learn about each other's strengths, and it also halts the momentum that originates out of each planning session, if the subsequent meeting is six months or a year later. All these reasons pose difficulty for the board to follow through and implement any of the activities that they propose.

A highly functioning board can help make STEM topics visible in the community to put it on people's radar to make it a priority. Being actively involved in the community will also help them recruit support from business leaders, who the majority of the interviewees recognize as the key people to help the TQE initiative. According to them, businesses drive the community priorities, because they are the ones who hire. Thus, business leaders are in the best position to inform the board of how it can improve. Other key people that they cited are community leaders, local politicians including the mayor and county commissioners, university personnel (directors, provost), and teachers.

The board has future activities planned to accomplish their mission of improving the quality of STEM education in the community. They planned a summit (conference) to advocate and bring publicity to STEM issues in the community. Also they will use various platforms to advocate for teachers and better resources in the classroom. They will try to find money to allocate to teachers as incentives for teaching STEM subjects and perhaps scholarships for them to take courses to improve their content knowledge.

Executive Board and STEM Issues from the Director's Perspective

This section presents the director's vision and perspective of the TQE grant Executive Board, along with perceptions of STEM issues in the urban community. Specifically discussed are the director's view of the importance of STEM education and the constraints pertaining to STEM education in community. With respect to the current functioning of the Executive Board, the director discusses her expectations of the board members along with the challenges to effectiveness.

Importance and Constraints of STEM Education in Community

Being a native of the community, the director of the TQE grant and its Executive Board is familiar with the local contextual issues surrounding education, including STEM education. The director has resided in the community her entire life, and has been a professor at the university since 1989. Paralleling the beliefs of the other board members, the director recognizes the importance of STEM education particularly with regard to its seminal role in promoting economic growth both locally and nationally. The director also identified similar constraints to having a high quality of STEM education in the community, including inadequate preparation of K-12 teachers in STEM subjects and ineffective pedagogical approaches to teaching STEM subjects, lack of student motivation, little parental involvement, and lack of public awareness of the necessity of STEM education. She comments that, "I am concerned that despite a lot of federal funding since the 1950's, teachers in general still do not teach by inquiry or problem solving." The outcome of ineffective teaching results in a large number of educated citizens who lack scientific literacy. This compromises their ability to effectively participate in a knowledge-based economy that requires strong math and science skills. Specific to the community, she states that "[our state] is below the national average in college educated students, and [our city] is below the [state] average."

In addition, she also illuminates the issue of political divisions that are present, which adversely affect the quality of K-12 education in the city's schools. "The biggest bottlenecks in my opinion are related to bureaucracy in the school system, rather than focusing on kids and high quality teaching." She elaborates that there are several political issues, including the division between administration and unions at the K-12 level. At the college level, some of the administrators did not support math and science education as much as other subject areas. At the state level, the Democrats and Republicans have very different agendas, each of them advocating for their own party politics, which may or may not support public education.

Vision and Perspective of the TQE Executive Board Director

To address all of the previously mentioned impediments to having an adequate level of STEM education in the city, the director recognized that it would require a multifaceted approach to effectively change the current culture of education. Seeking to be a catalyst of systemic change for improving and sustaining a higher level of STEM education in the city, the director believed that it was necessary to engage a myriad of local constituents to advocate and provide the necessary resources. She acknowledged that the key personnel to do so are teachers and administrators of K-16 schools, businesses, the government, as well as parents and other prominent community members. To represent these diverse set of perspectives, the director engaged the TQE grant Executive Board in discussions of these issues.

The director's vision was informed by national reports, such as *Raising Above the Gathering Storm* (2007), that were adapted to address the specific local issues. "Many people do not know how to get started in the city. There are strong histories of failed projects that started, but ran into roadblocks in improving the K-12 schools." Recognizing the contextual challenges, the director's purpose for the TQE grant Executive Board ended up going beyond the requirements of overseeing the grant. She extended the vision of the Executive Board to incorporate the members' goal of initiating a cultural shift in the city to focus on STEM education in K-16. Admitting that, "it will take thirty years or more to really change a culture," the director wanted the Executive Board to be an initiator of an organizational change via a holistic approach. Rather than targeting one element at a time, the director notes the importance of addressing all constituents simultaneously. This included, but was not limited to, getting grants, improving teacher-preparation programs, increasing the number of graduates from STEM fields, developing public service announcements, creating advocacy groups such as the Executive Board, and bringing media attention to STEM education to put it on the political agenda.

The director's expectation of the board members' roles was that they would be actively engaged in advocacy-oriented activities, as after all, they were the ones who wanted to go this direction instead of the original grant design of overseeing the grant. The director thought the group would work collaboratively "for the good of the region rather than their own entity." She specified that she did not want individual members to "play tuft games for their own side or their own agency." To fulfill her role as the director and to keep the Executive Board functioning effectively, the director put pressure on the board members through phone calls and emails to encourage them to engage in advocacy activities. The director also communicated the board's activities to other influential community members, such as the president and provost of the university and the state senator.

From the director's perspective, the board did not function up to her expectations. Although the board had a lot of potential to draw upon each member's individual talents, she admits that they did not accomplished as much as she had hoped because of the difficulty in "getting people to do something outside of the meeting times. Everyone is busy in their lives and it is easy to go on without anything happening." Thus, the director acknowledges the importance of having a collective desire to take action in order for the board to be successful. "If people do not feel an urgent need to change, they will not change." The director expected the members to take initiative as leaders in the city to impact STEM, but the members always wanted specific directions about what they should do. Sometimes it seemed to the director that they were waiting for permission to do something instead of being leaders to begin initiatives. Many times, the members would generate ideas about what the director could do, but the director wasn't looking for more work to do. Rather, she wanted the members of the Executive Board to do the work of the board.

Additionally, some members of the Executive Board were unable to put their own turf issues aside for the sake of STEM education in the community. At one important meeting scheduled to plan advocacy strategies to impact the state budget in support of STEM activities, a local K-12 administrator sabotaged the meeting with complaints unrelated to the topic of conversation that day. Later, the K-12 administrator admitted that he was just upset with a university administrator and took it out at the meeting. At another meeting, a university administrator sidetracked the meeting by stating that early childhood and democratic education were his biggest goals; not STEM education. Members of the school district argued with him that STEM education was more important for the city's economic prosperity and the purpose of this Executive Board. These two scenarios serve as examples of the inability of members to set aside personal agendas for the improvement of STEM education in the community.

The Meeting Minutes and the Accomplishments of the Executive Board

We used document analysis of the meeting minutes to determine how the interaction and engagement of the board members impacted the function and implementation of the goals stated by the Executive Board. This section is organized in two parts. First, we present the findings from the meeting minutes. Second, we transition to a discussion of the accomplishments of the board.

Analysis of the Meeting Minutes

The Executive Board was formed in January 2006, after the grant had been running for one year. As stated in the grant proposal, the Executive Board met three times during 2006, in April, June, and September. The April meeting centered on discussions of identifying the issues and barriers to effective STEM education in the community. They categorized the issues according to the school-based problems, lack of community and parental involvement, and lack of business partnerships with the public schools and universities. The proposed concerns reflect the common concerns that the individual respondents reported in the interviews, as discussed in previous sections. The agreement was to create a synergy among these three categories. As the first step to achieving this goal, they focused their discussion on identifying and leveraging local resources at the school, community, and business levels. For example the Dana Corporation, a large manufacturing company with headquarters in the city and a robotics division, was cited as being a potential resource for high schools as a way to provide the students with hands-on experiences in science. The group agreed that separate resources exist in the community, and it would become part of the Executive Board's mission to bring them together and work collectively. Thus, the meeting concluded with the members agreeing to each brainstorm, and then proposed ideas were emailed to the group about the resources that can be utilized to promote the advancement of STEM education in the community.

The follow-up meetings in June and September were integral for the defining of the Executive Board's missions. In the June meeting, the members decided that the board would primarily engage in activities that were focused on advocacy and being a catalyst for improving STEM education in the community. They cautioned against being a board that echoed rhetoric without action and proposed ideas for engaging in practical activities to support their mission. To support the advocacy stance, they stressed the importance of being connected with media outlets to publicly advocate for STEM-related issues. They also proposed the idea of building relationships with local business and nonprofit organizations, such as initiating after-school STEM-related programs with the local Young Men's Christian Association (YMCA) or Boys and Girls Club. As being a catalyst for STEM education in the community, they decided to organize a regional summit to bring attention to the importance of STEM as well as the inhibitors to STEM education in the community. In the September meeting, they further discussed the conference, deciding that it would be geared toward K-16 educators, community members, and business representatives. Noting the importance of the role that business plays in the community, they strategized ways to encourage business representatives to attend the summit, since their presence on the board was limited. The minutes reveal that the board members felt as though the local businesses adhered to the belief that schools are educating for the 1950s, not the twenty-first century, and that their participation in advocacy efforts was key to success. Thus, they wanted a high representation of businesses leaders in the community to attend the conference and engage in dialogue with educators and community members committed to improving STEM education.

It is crucial to mention that the board members agreed about the infrequency of the meetings, and they suggested that meeting monthly would be most appropriate to achieve their stated goals. There was a mismatch between the Executive Board's desire to meet monthly to pursue roles of advocacy and catalyst for STEM and the grant proposal's two times a year role of grant oversight and fund raising to continue the program after federal funding ended.

The whole-group meetings that occurred in 2007 focused primarily on organizing the summit that was planned for February, 2008. They discussed the purpose, structure, and desired outcomes for the summit. They also focused on identifying potential keynote speakers, topics and facilitators for breakout sessions, as well as the inclusion of attendees who have economic, business, and/or political power in the community and the K-16 educational system. However, smaller groups of the board met monthly as they planned the summit.

The last Executive Board meeting took place in February, 2009, and it was more politically oriented than the previous meetings. Specifically, they discussed the governor's plan for reforming the state's education system for the twenty-first century and the Obama-Biden Education Plan and strategized ways to align their activities with these initiated policy recommendations.

Accomplishments of the Executive Board

This section documents the accomplishments of the TQE Executive Board with regard to advancing STEM education in the community. Specifically, the Executive Board:

- oversaw the writing of a Saturday essay, which was published in the city's newspaper about the importance of STEM education;
- helped plan the summit on STEM education, which had over 108 participants;
- helped create six action groups (an outcome of the STEM summit). These groups met monthly for a year and focused on:
  - Strengthening partnerships and relationships
  - Placement of teachers within school districts and STEM graduates within the region
  - Accountability for learning
  - Increasing the impact of instruction
  - Generating awareness and advocacy
  - Recruitment and retention of a talented STEM and STEM education pipeline

Because this paper is not solely focused on the nature of these accomplishments, we do not go into details about the work of the six groups. However, the university continues to push a STEM agenda to date, and STEM and STEM education are now part of the university strategic plan. Individuals also participated in another regional conference on advocacy in May 2010, which was broadcast to several counties through two local public television stations.

Discussion

The purpose of this study was to describe factors that are associated with a grant-funded Executive Board's ability to create the private, public, and political will to improve STEM education and impact policy in the community. More specifically, this study examined the following topics: The factors that influence the evolution and implementation of a regional strategy for STEM education, and the creation of the private, public, and political will to improve STEM education.

Research shows that the likelihood for success of any form of decision-making entity is increased when the top management team members' views align with those of the leader (Hitt, Ireland, and Hoskisson, 2008). This was evident in the TQE board, as the board members' views were aligned with those of the director. However, the success of an organizational-change board is also dependent on several factors other than the sharing of similar beliefs. The most common factors that increase the possibility for success include clear understanding and consensus of the board's goals and objectives, members' active involvement in all decision-making processes, and directors having strong leadership skills as well as adhering to some form of organizational framework (Houle, 1989; Green, and Griesinger, 1996; Hough, 2006; Prybil, 2006).

In addition to a lack of understanding of the purpose of the board, the primary inhibitor to the success of the TQE was the different perspectives of board members' roles. The findings strongly suggest that the board members perceived themselves and their roles as advisory in nature. As such, they assumed a passive role in assisting the director in her vision. They attended the meetings, they gave their opinions about various issues, gave suggestions about what she could do, and they provided resources when necessary. However, they did not, and did not feel that they were expected to, serve in any leadership capacity. Their perspective conflicted with the director's expectations of the board members. The director stated that she purposefully chose an eclectic group of individuals based on their prominent roles in the community. Each board member was invited to participate on the board due to their stellar leadership capabilities in addition to their deep understanding of the importance and issues surrounding STEM education both locally and nationally. This led the director to incorrectly assume that the board would automatically function as a collective entity, with each board member being self-reliant, taking responsibility, and being committed to the board's objectives. As such, the director expected the board members to draw upon their strengths to provide guidance and initiate strategic planning as necessary. Contrary to expectation, the outcome witnessed the director being the sole team member who adopted any form of leadership role, while the 26 board members only offered their assistance. Our findings suggest that the functioning of the TQE could have been improved if the members' roles and expectations had been clarified from the beginning. In doing so, each board member could have been actively involved in decision-making processes throughout the existence of the TQE.

This misunderstanding may have derived from the limited number of leadership roles that the director assumed. Effective leaders play the roles of an advisor, expert, mediator, organizer, explainer, discussion leader, interpreter, reinforce, spokesperson, intermediary, summarizer, and team builder. Our findings demonstrate that the TQE director practiced only a few of these roles, most notably expert, discussion leader, advisor, and spokesperson (Ingley and Van der Walt, 2003; Schmid, 2006). In addition, Bass and Riggio (2006) advice that directors should assign leadership tasks to each group member as a means to effectively develop within-group leadership. Doing so could have prevented the disparity between the director's expectations and the board members' perceptions of their roles.

The use of an organizational framework may have benefitted the TQE, as it would have encouraged the director to effectively communicate the board's goals, coordinate the group's efforts, and keep the group focused on the board's primary objectives (Nicholson and Kiel, 2004; Pye and Pettigrew, 2005). Leavitt and Bahrami (1988) offer a framework that emphasizes the importance of structure, information and control methods, people, and task. Within this framework, they assert that an organization's structure is shaped by the tasks to be accomplished, but mostly by the people involved.

The composition of the leadership team, particularly their interpersonal relations, highly influences the outcome of change efforts (Forbes and Milliken, 1999; Adobor, 2004; Li and Hambrick, 2005; Stewart, Fulmer, and Barrick, 2005). It is also possible that there were preexisting interpersonal relationships among the board members that made it impossible for them to work together effectively. For example, there were union and administrative representatives who had adversarial viewpoints, business leaders who had previous negative experiences working with the local school district to get them to improve, and a university administrator who had previously expressed his unwillingness to support STEM-education efforts over his own interests.

Organizational Change and Leadership Dynamics

The majority of organizational-change efforts fail (Ford and Ford, 2010). Of the few that are successful, effectively sustaining the changes over time is an extremely difficult task (Hargreaves and Fink, 2003). With specific regard to educational reform, change efforts must navigate complex and ever-changing school and community contexts. The shared governance structure of leadership teams strongly influences the leadership capacity of the team (Scribner et al., 2007), and the success of reform attempts is most often attributed to the dynamics of the leadership team. We use cultural exchange theory as a framework to examine the collective efficacy of the TQE, and to examine how the board members functioned as a team.

Cultural exchange theory is an appropriate framework to guide the discussion of the functionality of the TQE, as it was comprised of members who represented diverse professional backgrounds, each of which has its own cultural system. Cultural exchange theory "describes a transaction of knowledge, attitudes, and practices that occurs when [individuals] representing diverse cultural systems . . . interact and engage in a process of debate and compromise" (Lindamer et al., 2008, p. 236). Drawing upon the tenets of this theory that stresses the importance of collaboration through the negotiated processes of communication and the sharing of knowledge (Brekke, Ell, and Palinkas, 2007); successful and sustainable organizational-change demands the establishment of a collective group culture. The formation of this organizational culture necessitates the practice of stakeholders stepping outside of their comfort zone to traverse unfamiliar territory in an attempt to engage in collaborative forms of group interaction (Willging et al., 2007) that draws upon the strengths of each group member.

Collective efficacy refers to the shared beliefs about a group's abilities to perform specific tasks (Bandura, 1997). Research consistently demonstrates that a positive correlation exists between level of collective efficacy and team performance (Gully et al., 2002, Stajkovic, Lee, and Nyberg, 2009). Factors that influence the level of collective efficacy include the task-relevant knowledge of each individual, the teamwork behavior ability of the individual, and initial team perception of success; the greater the task-relevant knowledge and perception of early team success, the better the overall performance of the team (Tasa, Taggar, and Seijts, 2007). Collective efficacy is a product of the team members' interactions, therefore, it emerges as a group-level attribute, rather than equaling the sum of each member's perceived self-efficacies (Bandura, 1997; 2000). The failure of the TQE members to demonstrate collective efficacy and to actively engage in shared knowledge transfer and to assume roles that were different than they had been used to, denied the board the possibility of enacting any form of sustainable reform practices.

Recommendations

Despite its common use and documented success in the business sector, community psychology and social work, organizational-change boards (a board explicitly selected to implement change in a given organization) are rarely used in the field of education. However, with increased attention and efforts directed toward educational reform, it is anticipated that organizational-change boards will become ever more common in education. Thus, foreshadowing an increase in their use while recognizing the limited literature on executive boards in relation to issues surrounding education, we use what we have learnt from this study to offer suggestions. Drawing upon research for barriers to change, we provide the following recommendations for individuals who are interesting in establishing an education-focused organizational-change board:

1. Purposefully select members based on their qualifications and ability to work as a team with strong collective efficacy (Bandura, 1997; Stevens and Campion, 1994);
2. Clearly communicate the purpose and goals of the board to all of the members at the time of board establishment, and continually remind them throughout the meetings (Stiles and Taylor, 2001; Pye and Pettigrew, 2005; Stewart, Fulmer, and Barrick, 2005);
3. Explicitly clarify the roles of each member (Aguilera, 2005; Huse, 2005);
4. Establish norms of acceptable behavior (Forbes and Milliken, 1999; Ingley and Van der Walt, 2003; Nadler, 2004);
5. Meet regularly and often enough to establish a group culture within which the members can identify as a collective and function as a group (Coulson-Thomas, 1991; Charan, 1998; Forbes and Milliken, 1999; Conger, Lawler, and Finegold, 2001; Ingley and Van der Walt, 2003; Sundaramurthy and Lewis, 2003);
6. Involve all members in decision-making processes. Stufflebeam et al. (1971) assert that there are four stages involved: awareness, design, choice, and action;
7. Director should encourage leadership development within the group by giving leadership tasks to each group member for which they are responsible (Forbes and Milliken, 1999; Ingley and Van der Walt, 2003; Aguilera, 2005; Huse, 2005);
8. Director should assume the characteristics of leaders, including being an advisor, expert, mediator, organizer, explainer, discussion leader, interpreter, reinforce, spokesperson, intermediary, summarizer, and team builder (Schmid, 2006; Ingley and Van der Walt, 2003).

NOTE

1. Funded by the U.S. Department of Education through the Teacher Quality Enhancement Grants program, Title II of the Higher Education Amendments of 1998.

Chapter 7

STEM Stakeholder Response to Enacted Educational Policy

Carla C. Johnson
and Virginia L. J. Bolshakova

The United States is in the midst of a formidable challenge as the pipeline of STEM (Science, Technology, Engineering, and Mathematics) expertise continues to decline. Policy makers and national leadership have argued the "future prosperity of the United States" can only be preserved through the strengthening of science and mathematics teaching and learning in public schools (Committee on Prospering in the Global Economy of the 21st Century, 2007). There have been an emergence of commissioned reports related to the STEM talent crisis including *Rising Above the Gathering Storm* (2007) that proposed the following: (1) increase of the talent pool through improving K-12 science and mathematics education, (2) sustain and increase long-term basic research related to the economy, security, and quality of life, (3) increase the attractiveness of the United States to recruit and retain the best and brightest scientists and engineers in the world, and (4) increase incentives for innovation (Committee on Prospering in the Global Economy of the 21st Century, 2007).

In addition to the *Rising Above the Gathering Storm* report published in 2007, others have argued that STEM skills are the key to future success for all students working in the twenty-first century (Sanders and Wright, 2008). There is a plethora of data that indicate that the majority of highest-paying jobs will require mastery of science and mathematics skills, as one of every three jobs by 2015 will be STEM related

(National Science Board, 2007). It is clear that mastery of science and mathematics concepts are necessary and correlated to success in college, to economic growth and development, to national security, to innovation, to entrepreneurship, and to competitiveness in the global market (i.e., U.S. Commission on National Security/21st Century, 2001; Business Roundtable, 2005; Committee on Science, Engineering, and Public Policy [CSEPP], 2007). The Carnegie Foundation report (2009) entitled *Opportunity Equation* recommended four priority areas for moving the United States forward in the global arena, which included: (1) Achieving higher levels of mathematics and science learning for all students; (2) Establishing common standards in mathematics and science that are fewer, clearer, and more highly coupled with aligned assessments; (3) Supporting improved teaching and professional learning; and (4) Creating innovative, new schools and systems that are more effective in preparing students in mathematics and science. Moreover, it is clear that the future of the United States as a prominent leader in the global arena depends on deliberate and responsive change.

Issues with K-12 STEM Education

Educational policy, more specifically the No Child Left Behind (NCLB) Act of 2001, has worked to divert academic focus in K-12 schools away from teaching science and mathematics in authentic, real-world contexts. The reality of mandated assessment of mathematics and reading, as well as ambitious goals for reaching adequate yearly progress (AYP) for schools has created insurmountable barriers to teaching science and has taken away the ability for teacher creativity (Ruby, 2006; Johnson, 2007a). As Linda Darling-Hammond so eloquently stated in a recent article in the *Chronicle of Higher Education* entitled, "We need to invest in math and science teachers," she shared, "The shortage of well-trained, career-committed math and science teachers has created a viscous cycle for our nation: Because those subjects have frequently not been well taught, we have a nation of math-phobic citizens, few of whom are prepared to pursue higher-level math and science in college" (2007, p. B20).

Legislation that was intended to level the playing field for urban schools has worked to place them further behind, complicating the realities of urban schools that often lack strong building leadership, resources, and teacher knowledge of the effective practices needed to reach all students (Anyon, 2001; Barton, 2001; Lynch et al., 2005; Ruby, 2006). Teacher quality is a fundamental problem in the United

States for several reasons. Many science teachers feel that they did not receive adequate content knowledge or pedagogy in their university preparation to teach science effectively (Crawford, 2000; Keys and Bryan, 2000; Wright and Wright, 2000). In urban schools, often teachers face challenging environments with little support and rely on teacher-controlled instruction due to management issues, limited resources, and other barriers. An opportunity gap has been reinforced indirectly through policy, as children are not receiving the opportunity to learn science and mathematics in highly effective classrooms (Anyon, 2001), which has resulted in a persisting achievement gap for children from under-represented groups (Stoddart et al., 2002; Johnson, Kahle, and Fargo, 2007a). As Barton (2001) argued, "There is a critical need to address the particular science education experiences of urban youth, especially those from marginalized communities."

As Wallis (2006) clearly articulated in his *TIME* magazine cover story, "How to Bring our Schools out of the 20th Century," there is a pressing need to reform how we are preparing our future generation to better reflect the skills they will need to compete in a global society, stating, "Teachers need not fear that they will be made obsolete." Wallis (2006) shared that "the nation's more forward thinking schools are bringing this back," referring to the focus on more than content—on preparing students for an interdisciplinary, project-based, technological society.

In Ohio, many promising students fail to enter STEM careers for a variety of reasons. Some lack knowledge of the opportunities available in STEM fields. Many are daunted by the perceived financial costs of a college education. Others have not taken the appropriate high school courses to prepare them for higher education. Even when students enroll in STEM pathways in college, many leave the field or drop out of college. Today, only 32 percent of students graduating from Ohio high schools are prepared for college. This percentage ranks Ohio twenty-seventh among the 50 states (Achieve, Inc., 2007). The Achieve, Inc. report, *Creating a World–Class Education System in Ohio* (2007), postulates that "making Ohio's educational system exceptional is the one action that would most improve Ohioans' standard–of–living" (p. 3).

This study will provide a lens into the beginning steps of STEM educational reform in a region of Ohio that has experienced an economic decline along with the decline of the U.S. automobile industry. The research questions that were the focus of this study included: (1) What are the needs that STEM stakeholders identify that should be addressed to move toward improving the pipeline of STEM talent in their region? and (2) What are the strategies that STEM stakeholders propose and/or develop to address the identified needs?

Theoretical Framework

Kurt Lewin's (1947) change theory has been widely used in organizational- and educational-change work over the past several decades. This three-step change theory is based upon the premise that behavior is guided by a dynamic balance of forces working in opposite directions. According to Lewin (1947) driving forces move stakeholders toward change and restraining forces work against reform.

The three-step model is used to navigate these forces and move collective groups toward change. The first step in moving toward change is to unfreeze the status quo or existing situation. This is achieved by either an increase of driving forces and/or a decrease in the restraining forces. Robbins (2003) proposed this could be accomplished by motivating stakeholders, building trust, and revealing the need for change through collective brainstorming of needs and potential solutions. The second step of the Lewin (1947) model is movement. In this step, the system moves toward the desired changes during what is considered the period of transition. The development of new perspectives and strategies would take place in this step and leveraging partnerships and leadership to support the change. In the third step, refreezing occurs when the system has reached desired change, and supports are in place for the changes to be sustained. Without the implementation of this step, reforms and changes may disappear. In this step reform is institutionalized. This may include the integration of new strategies, programs, and partnerships within the context of the system as routines.

More recent work in change theory has built upon the work of Lewin, including Fullan (2006) who includes seven principles that mirror Lewin's model including: changing the context (unfreezing), motivation, capacity building with focus on results (leadership focused on needs/solutions), learning in context, reflective action, tri-level engagement (leveraging partnerships to support change), and persistence and flexibility in staying the course (support for changes to be sustained, institutionalized). Fullan (2006) argues for motivation to take place there must be "moral purpose, capacity, resources, peer and leadership support, and identity" (p. 6). Capacity building, according to Fullan (2006), is "any strategy that increases the collective effectiveness of a group to raise the bar and close the student gap of learning" (p. 7). Additionally, reform must have the capacity to change the larger context (Fullan, 2006). Reflective practice includes "shared vision and ownership" and "behavior changes to a certain extent before beliefs" (p. 7). Tri-level engagement includes school and community, district, and state, as key to systemic change (Fullan, 2006).

Community-based reform can be powerful and has received increased attention within the last ten years (Shirley, 2009). The work of Lewin and Fullan will be used as the theoretical frame for this study and findings will be used to further advance our understandings of change theory.

Methods

Setting and Participants

The participants in this collective case study (Glesne, 2006) were participants in focus groups and a regional STEM conference in the northwest Ohio region during 2007–08. The focus group sessions were conducted in the fall of 2007, which included a K-12 teacher group (12 participants), a K-12 administrator group (16 participants), a business/community group (26 participants), and a higher-education group (21 participants). The regional STEM conference took place in February 2008, and there were approximately 120 participants at the event with a breakdown of subgroups much like the focus-group sessions.

Participants in the focus group were invited to attend and discuss the current status of STEM in the region. The session began with the facilitator reading a brief paragraph from a local newspaper that was followed by semi-structured questions and group discussion of needs of the region from their own perspectives.

Participants at the regional STEM conference registered online and selected from one of five strands that reflected needs that emerged in themes across the fall focus groups that will be discussed in detail in the findings section. There were some speakers during the morning of the event, who discussed STEM from local and national perspectives. This was followed by three hours of breakout group sessions where the participants began discussing the collective needs from the focus groups and brainstorming strategies that may address some of those needs. The regional STEM conference provided breakout group participants the opportunity to join work groups that continued meeting monthly beyond the event to roll out some of the strategies and deliverables suggested by the groups. Approximately 50 percent of the participants in the breakout groups continued to participate in the follow-up work after the event.

Data Collection and Analysis

This qualitative study utilized multiple data sources that were triangulated providing convergent validation (Glesne, 2006). Data for this collective case study were gathered through a series of focus-group interviews with constituent groups, as well as transcribed meeting recordings, participant observation, and artifacts such as posters, agendas, and meeting minutes. Qualitative data were analyzed as they were collected and then again at the conclusion of data collection to identify trends across the study that represented the collective needs of the groups by the primary author of this study and two other faculty researchers for validity (Wolcott, 1994; Glesne, 2006). The focus-group sessions were two-hours long and followed a semi-structured protocol. Questions for each focus group are included in appendix 7.1.

Findings

The overall collective group needs were determined through a series of focus groups conducted with each prospective group including K-12 teachers, K-12 administrators, higher-education faculty, and business/community members. There were five themes that appeared across all four groups, which constituted the primary areas of concern. Within each of the five primary areas, more specifically detailed subcategories also emerged. Each of these areas and categories will be detailed and discussed in the following sections and some key thoughts from participants will be shared as examples of the extensive conversations that took place in each of the four, two-hour discussions.

Accountability for STEM Learning

Participants in all four focus groups shared their thoughts regarding the need to greatly increase accountability (holding schools and teachers responsible for ensuring students learn) for learning STEM subjects. More specifically, there was an overwhelming expression of concern across all four groups regarding the quality of STEM teaching in schools, the lack of science instruction, the under utilization of data to inform instruction, the weak alignment of STEM disciplines between high school and postsecondary institutions, and the overabundance of instructional time devoted to test preparation in K-12 schools.

Effective STEM Teaching

The quality of instruction in our nation's schools has been of grave concern for many years now. In this region, it was a reality that was difficult to ignore. Effective teaching includes engaging students in doing inquiry, explorations, utilizing problem solving and critical thinking about the world around them, as well as exploring their own questions. One participant's response highlighted the main concern for the group, sharing that, "there is no accountability in the system to make sure that teachers like that are either eliminated from the system or improve." The acceptance of mediocre teachers—often promulgated by protection of teachers' unions—had resulted in students moving through the system without mastery of basic skills.

> How can I have an eighth grader there who has been through eight different teachers, nine actually, and now in eighth grade he can't even add. You have to go back to the school system. A lot of people don't realize that when you get them in high school, some of them have missed eight years of foundation. You never start building a house from the top down. You have never seen them put a roof on a house with no foundation. So you have to have a foundation to get where you need to be. (Administrator focus group).

The quality of instruction was further exacerbated by inconsistency within the schools, as one higher-education participant argued, "There is a big disconnect between what kids are learning—in whether it is being taught, not taught or whatever." For reasons such as lack of confidence teaching various subjects, to more instructional time devoted to test preparation and focus on reading and mathematics, some subjects received less emphasis and often were omitted from the school day.

> There is a big difference between everybody teaching the standards, covering the standards like they are supposed to do and the students actually learning and mastering. We should be at the point now where everybody is teaching it... there shouldn't be teachers out there that are not following the standards—but for them learning it, that is a different story. (Teacher focus group).

Participants from the business and community stakeholder group interestingly had a good understanding of what should be taking place in the K-12 classrooms in the region. As one community member asked, "How many of us teach our kids by putting them in rows with a blackboard behind us?" The question was followed by conversation regarding authentic learning experiences, collaboration, and opportunities to engage with the natural world and share with others as ways that learning takes place outside the classroom and often at home.

A discussion was led by members of the business community about global competitiveness and educational systems abroad and the reality that the regional educational system was failing to meet the needs of their students.

> China and India for example, have more honors students than we have students. If you just go by statistics and population. And they have been investing heavily in science and technology education over the past twenty to thirty years, whereas, we are playing video games. You know, the basics need to be taught to them, but our students don't have... our urban students, our inner-city students don't have the exposure.

This comment was followed up by a question from a community member, "So, is it really the students who are failing or is it the instruction that has failed?"

Routine Teaching of Science

There was much discontent from stakeholder groups regarding the lack of regular science instruction in regional elementary schools. Many participants failed to understand how schools have been allowed to dismiss the teaching of such a critical discipline. One administrator shared that his district had, "all but stopped doing science in the elementary." The results of the decision to minimize science were not surprising. A teacher participant shared his frustration regarding the lack of minimal background experiences and knowledge for his high school students.

> Our biggest problem is the elementary, I hate to pick on them and my wife is one of them, but she sees it. A lot of those teachers don't teach the math and science because they are not comfortable with it and then they get to junior high where they actually have a math and science teacher and you expect things and they just don't have the background in it.... Of course we get it in high school too. It is like their education in math and science starts at junior high.

There was a collective desire across the four stakeholder groups to begin holding schools accountable for teaching science, and ensuring that students are receiving high-quality instruction in all disciplines as one community participant clearly articulated, "We have to start holding teachers accountable to do the right thing and give them support to do the right thing." Further conversation revealed that the problem may reside outside of the teacher's control. Some districts had not included science in their district curriculum frameworks due to giving priority to reading and mathematics with the hope that more focus on these areas may result in more progress toward AYP for struggling elementary schools.

> And to be truthful, our elementary science has not been mapped, and that is because of the focus on reading and math. The standards are still there, it is just that the mapping piece is not in there. But seventh, eighth through high school, even the mapping piece is in there. (Administrator focus group).

These sentiments were echoed by the K-12 teacher group and one participant in particular was very cognizant of the long-term impact of those decisions on student learning.

> We need to make sure that math and science are being taught every day, especially at the elementary level. If I don't like math or science, I am not going to teach it, henceforth, I shut my kids out of those experiences. When they get to seventh or eighth grade they are expected to take it every day, but they haven't gotten the basics of the foundations in first through sixth grade. We as educators have done them in before they got started.

Members of the business and community stakeholder group echoed concerns of the other groups and realized that the lack of science experiences results in lack of motivation to learn science in high school.

> I think before they get to the high school they don't do science much. They probably read about it a lot. So then you wonder by the time they get to high school why they don't like science—well, probably the same reason they don't like social studies. All they do is read about it here and there and they have had few opportunities to do science.

The discussion of this topic sparked an outpouring of emotion from the business and community group as they passionately took ownership of this issue. One participant stated it best, who proclaimed, "We have to do something about TPS not getting science to the children in the school."

Data-Driven Instruction

A need emerged in the area of using the data that is collected from the various benchmark and state-level assessments to inform and drive instruction. Many stakeholders in the business and community groups, as well as some from higher education, believed that this was already taking place and were surprised that the wealth of data that was being collected had not been put to better use. This conversation started in the business/community group as one participant ignited the topic through the lens of the students who are asked to jump through hoops though they receive nothing more than negative reinforcement in many instances, as the data is not used to provide support for student learning.

> All we do is test and kids are getting so discouraged. We are using it more as tools of failure. They see that they didn't pass the OGT. Do we actually use these results? I know that the teachers get them back, but are we using these? So I think the student, by the time they get to the high school level or even in the junior high and maybe the upper levels of elementary—they are so frustrated because they have gone from kindergarten up to fifth grade, haven't met the standards and skills that were needed to be met in those grades. They are starting to get frustrated so they are getting the attitude that "I don't care." They are not getting support at home, now they are failing in school because they haven't met the standards. We are testing them like crazy! They are taking those tests and once they see the failure they are getting discouraged.

The participants in the administrator group echoed these concerns, as one participant discussed, "This is how I see it. Another test, here we go again. But the problem is that we are not using any of these results to help. Our students, they take these tests and never even see the results." The K-12 teacher group expressed that they have the desire to use the data but many have not been taught how to interpret the scores and have not been provided support to implement interventions during the school day. One of the teacher participants shared, "Our biggest problem that we have is that we haven't trained ourselves and trained each other in how to look at that and evaluate the results and ask ourselves, What do we need to do?"

Improved Alignment between High School and Postsecondary Institutions

Higher education institutions were also discussed in their roles of preparing future STEM teachers, as well as educating students in STEM and non-STEM disciplines. Alignment of high school and college course work was identified as an area for improvement as many students are faced with taking remedial course work at the university before starting college-level courses. This was discussed as an issue by all groups, and it was clear from the higher-education group that this is a source of frustration for all.

> It is really demoralizing for our faculty who teach those classes—and it is not great for students. The university is very concerned about retention

> rate, and this is hitting students right in the face. "I got A's in high school, why am I getting Cs, Ds and Fs at the university? What is wrong with me?" I am very concerned about that, and I just don't know quite what we might be able to do to address that, if anything.

One chemistry faculty member shared that 700 students were essentially retaking high school chemistry at the university this year because they could not pass the placement test to get into general chemistry. A biology faculty member shared that the failure rate in introductory biology classes is 30–40 percent at the same institution. The sense of frustration was clear, as another higher-education group member shared, "Students can walk in the day after they graduate from high school, come into our classes and they are not prepared."

Many of the focus-group participants saw the alignment between regional high schools and the university as a huge need.

Avoid Allowing "the Test" to Drive Instruction

All of the focus-group participants agreed that accountability for student learning was an overall area of extreme importance. However, the method that the federal and state governments have employed to determine progress of students in K-12 schools has often run counterintuitive to the goal of holding schools accountable. One good example of this that was discussed in detail by the focus groups is the use of reading and mathematics performance to determine school AYP. A participant from the administrative group defended his district's decision to focus more instructional time on those areas due to the emphasis of the assessment, as he stated, "I think that we should realize that there is a national system called No Child Left Behind that is driving the academic achievement test and science does not appear there until later." This person clarified that he personally does not support this decision and emphases away from science and other disciplines.

Many participants also discussed the indirect impact that high-stakes assessment has had on routine practice within schools. Struggling schools and those aspiring to reach their individual AYP goals devote classroom time to test preparation and review of concepts for weeks at a time. A member of the business/community group shared her thoughts on this practice.

> The institution of mandatory state-wide exams has greatly influenced how things are taught. My daughter is in the eleventh grade now. This has been phased in her lifetime as a student and she says they spend at least half the time studying for the test. They don't care if you learn anything you just have to study for the test and they take all these practice tests and they are

> used to doing this kind of examination. I think this has an enormous effect on our schools as we see it published in the papers of how various schools did. So how do we recognize the impact of the mandatory state testing and what does that mean for what we want to accomplish?

Unfortunately, many of the K-12 teacher participants in the focus group had only a vague understanding of how to interpret state assessments, as they had only been given unclear mandates and increasing pressure to improve in designated areas. Teacher participants shared that this obviously had a great impact on what they choose to do and what they do not do in their classrooms.

> Our principal told us we have to raise math scores. We have to raise performance index and attendance and we have no idea what our performance index is. What causes it to be where it is now? What needs help? We have none of that.

The higher-education focus group expressed concern and frustration with the shift that has occurred within K-12 classrooms and the associated emphasis on test preparation. As one member of the group shared, this approach has resulted in curriculum being, "An inch deep, mile wide." There was also concern that multiple-choice assessments were so heavily emphasized, though only a snapshot in time and void of the ability to measure what students can actually do.

STEM Awareness and Advocacy

Parental and Community Knowledge and Engagement

One of the biggest concerns of stakeholders was that parents of are not aware of the importance of STEM skills to the future success of their children in the twenty-first-century job market. The emergence of STEM emphasis coming from policy in response to the declining performance of the United States globally in science and mathematics achievement is something that has not been widely disseminated to stakeholders—including parents. The lack of STEM awareness and advocacy were identified as areas of concern for all stakeholder groups.

As one community group member shared, "I talk to families and they are looking to the schools to tell them what they need to do." The four groups discussed the trust that parents place in the school system to provide an education for their children and to reach out to them as partners in the system. There was also discussion of how the educational system has changed over the past two decades and that now, more than ever, parents are needed at the table as part of the team to ensure that their children are successful. A K-12 teacher from the teacher focus group shared his thought that in order for students to be successful, "that comes down to parents. Your parents are your first teacher. Again, we can't leave the families out of the equation."

It was clear to the participants in the focus groups that there is a community educational piece that must be delivered for parents to be on the same page with teachers and other stakeholders in the STEM community. As one member of the higher-education group argued, "Sometimes we wrap ourselves up—even the acronym that we are using, STEM—if you went out on the streets and said STEM nobody would know what you are talking about." Participants in all groups were concerned that many parents of students in the urban schools in this region are lacking advanced degrees and have possibly not had strong foundations in STEM disciplines as well. A participant in the administrator group explained that this further substantiates the need for community awareness and potential educational programs on STEM.

> It is hard to have them to be advocates for STEM when they have no concept of an equation, have no concept of geometry, algebra. Some of the things you are asking them to advocate for if they are a dropout, what you are asking is for them to advocate something that they have no conceptual knowledge of.

Members of the business and community focus group shared concerns that were echoed by other groups regarding parental lack of understanding of the problem solving and critical thinking skills that are required to successfully compete for the good jobs today. One member of this group articulated the shift in the automobile manufacturing field that had served this region so well in the past. The jobs that parents may have had will not be the jobs for which their children will compete.

> It is also larger than what is happening in the K-12 schools though. I think a lot of parents still think that their kids can go through high school and get a job at GM [General Motors]. They think that they can go through high school and get the jobs that they got, and it is just not happening anymore. You have to have a four-year degree to be hired at GM or at least a two-year college degree.

Parental advocacy is critical, according to the members of the focus groups, as one administrator shared, "Parents have to have an interest in it as well." The larger concern is that parents are not well versed in the transformation that is taking place within the United States and in the region in particular. There is a crisis in the shortage of STEM talent and schools are failing to meet the needs of students in these areas. All of the focus groups discussed that this may not have trickled down to all parents. One teacher shared her own experience with attempting to educate parents of the issue.

> So when we met with the parents, you know that these parents weren't even aware of the crisis that is going on within the school. The school is in academic emergency. They had no idea what that meant. They had no idea of the high drop-out rate in our public school system.

STEM as Economic Development

The quality of life within this region has declined and more specifically within this urban area there have been many families and businesses that have chosen to move elsewhere. The decline of the automobile industry initiated this movement. However, the "brain drain," as many from this region call it, has further exacerbated the situation. The STEM talent pool in northwest Ohio is shallow, and this has resulted in only marginal attraction of new industry. Many of the participants in our focus groups see the STEM crisis as an issue of economic vitality, as one participant of the business group shared, "One of the things maybe we should start doing is looking at this education crisis more as an economic developmental issue. I think a lot of our culture thinks of it as a social issue, and it really isn't."

One way of beginning to address this crisis and the direct impact on regional economy, according to many participants in the focus groups was articulated best by a higher-education stakeholder who stated, "We need to begin to think about this (STEM) as something that is fundamental to any job." Through ensuring that all students are prepared in STEM, the pipeline of STEM talent could be revitalized and participants believed that economic progress would occur. The groups gleaned that the result of no action would be the continued loss of economic opportunity for all, as one member of the business/community group argued, "As a country if we don't get the situation with our education piece together, they will have no choice but to look at Asia when they look at moving all their companies. Do our people understand that we are not going to be able to compete in the next twenty-five, fifty years in the global market if we don't do something about this education system?" K-12 focus-group members concurred, as one teacher shared, "If you can't produce the workforce they are going to need, they are not going to come here."

Marketing STEM Awareness

STEM awareness was an area of concern for all participants in our focus groups. The business and community stakeholder group engaged in discussion of how to synergize the regional community behind the importance of STEM. One business leader argued for the need to take an active role and to market the STEM crisis as an opportunity for the region to move forward in a purposeful and positive way.

> We have been a quiet partner in this community for years and years. We are not so quiet any more. But we need to market this and we need to show it from a positive aspect. And just merely mentioning these things in the media make a difference.

Those who were part of the group who had seen reforms come and go in this region were a bit apprehensive to believing that things could in fact change. One higher-education leader shared, "We have no public will to make things better and hold schools accountable." Other members of focus groups agreed that the task would be difficult, but that it was not impossible.

> I don't know how we are going to do it, but I think if we get people in the community involved and parents involved, it could work. I can't think that we can find anyone who agrees with the way that it is going right now.

Many of the challenges that were shared were focused on the tightly controlled, low-performing public school systems. There had been previous attempts to build partnerships that were unsuccessful in transforming the system. One community-group participant shared, "Our local system is killing us, I think—in some respects." The teacher group shared a similar view and one teacher discussed, "I think that we are a victim of our own system. We decided in this country to do schools locally and therein lies the problem."

Each group decided that the first step of marketing the need for STEM may include the need to get the public more engaged in the K-12 schools on a regular basis. One administrator from a large public school system in the region shared thoughts about the need for improvement and the potential of STEM.

> They need to come in and see what is going on. This is the only way to get people to buy into what is going on. And they can't buy in unless they are here. When they see those kids sitting in those classrooms, they are going to say, this is not working.

Improving STEM Instruction

Funding, Resources, and Support

It is clear that effective schools have access to funding, resources, and support from leadership and community. This was identified by the four focus groups as an area of need for their region. The current funding structure in the United States is inequitable, as school systems in lower-tax-base areas do not have the resources that their suburban counterparts enjoy. A member of the administrator group shared, "The most important is easy to say, and hard to do, we need some funding. And I think that is a response that you will hear virtually from everyone is they don't have the budget to do this."

In terms of resource, the largest divide is in access to technology and this was discussed in detail by all groups, but particularly by the K-12 teacher group, as one member shared, "You talk about STEM and talk about trying to integrate math and science. The most basic that we need is to get them the access to even the most basic computer and computer skills and build upon that. Imagine if every kid in your classroom had a laptop." It was clear that technology was viewed as a luxury by this teacher in an almost unattainable view.

Support from leadership and community was an area of need discussed by all groups. The building leader can provide support in the way of time to teach, time to plan, and time to collaborate. One of the K-12 teacher-group participants shared personal experiences of the impact a strong leader could provide for a school.

> We have a great principal at our schools. Anything we want. He will find money if it will help the students and help their progress. Just having that support, having that backbone, knowing that you can go to that person and they will be there and they will back you up. Doing whatever they possibly can.

In addition to supporting the teachers with funding, resources, and moral support, the groups discussed the role of the principal as an instructional leader with a role in the curriculum implementation. A member of the business/community group shared thoughts about additional ways that building leaders could contribute to the success of the school.

> The role of the assistant principal should be looking for curriculum and resources. The principals should be supporting instruction in the classroom. They should be supporting their staff. They should be supporting their teachers in any way that they can.

Unfortunately, some teachers are not well positioned to take advantage of support opportunities from the community or from their institutions of higher education. One of the participants from the university group shared her frustrations with trying to form a partnership with a local classroom teacher.

> I actually asked the science teacher there if he would like some assistance and he said, 'You know, I appreciate the thought but you don't understand. I have to get fifty kids into this classroom and then I have to turn them around. I don't have the time to take the microscopes out and put them back away. So it is very superficial, so we just can't do anything exciting or creative in the classroom."

Innovativeness—Teaching for the Future

The reality of schools in this region was discussed by all groups and one member of the higher-education group did an excellent job of summing up the sentiments of the collective groups as he shared, "If Ben Franklin were to walk the earth today, the place he would feel most comfortable would be our public schools because they have changed the least." The grave importance of changing the way that schooling is conducted was discussed in great detail. It was clear that groups felt that schools were not preparing students to successfully contribute to society and/or secure a job of the future. As one teacher shared, "We cannot teach them the way that we were taught." The reality for many schools in this region and across the United States is that teachers and schools have failed to keep up with the changing needs of society. An administrator shared her frustration regarding the instruction in her school.

> Is that my teachers aren't totally aware because our population has changed and my teachers haven't changed. They are thirty, twenty years in and they don't understand that they can't teach the children in their class any more the way they used to... or do what they have done for twenty years.

Many groups discussed that the emphasis on rote memorization in preparation for high-stakes assessments have diminished individual teacher creativity. One community group member shared her observations that, "teachers have lost their ability to be creative. They feel like they can't go outside—that is what we are talking about, to be creative and teach them in different ways."

One of the most difficult realities was a topic of discourse for all groups, which was that things are changing rapidly within our society with innovation and technology, and it is difficult to know what the jobs of the future will be. It was agreed that skills such as critical thinking and problem solving will be essential to the jobs of the future.

> You know there are a lot of research and development departments in our companies also. I think we need to begin to teach that. Not only to get kids to understand that if we want to become again a country that is innovative and pioneering new efforts and new technology, research and development is going to be key to this country. (Higher-education group member)

Perhaps a business-group member stated the thoughts of the collective group best, "One clear factor you can take from that is that we will not be what we are today. So if you want to know what not to educate to—don't educate for today, because it will not be what we are today."

Reward Excellence

Stakeholder groups discussed the quality of STEM instruction that is currently taking place within the schools in the region. In addition, the issues of low teacher salaries and lack of recognition for effective teachers were debated. The groups decided that it would be essential to reward excellence in STEM teaching to promote change within the system. A business-community member shared that this model has worked in other careers and has produced quality through competition and high expectations.

> It is a matter of learning how to create a free market where what you pay is what you have to pay and then you begin to pay based on merit. Not just tenure and showing up and those kinds of things, because people are competing for those teaching jobs.

Members of the administrative group were enthusiastic about differentiated pay and one participant shared, "The reality is we have to pay teachers differently." In this region, highly qualified new teachers are turned away by districts who have systems established that favor teachers who may be teaching out of area or without highly qualified status, making it difficult to improve the quality of instruction in those schools.

Another potential way to reward excellence is through supports that administrators could provide. Members of the business group shared ideas of other incentives that could be used to reward the best of the best. Observations of a member of the business community were that teachers could be provided more time off to avoid burnout sharing, "Good teachers work very hard and are exhausted." Additionally, the teacher group shared that it is not enough in itself to reward teachers with pay for excellence.

Teachers need supportive building leadership as one teacher shared, "The differences between building climates and culture and leadership makes a difference in teacher's motivation to excel."

Improved Interventions

For most teachers, the school day provides few opportunities to individualize instruction. This issue was discussed by all groups as an area of need for all students to learn STEM subjects. Early identification of students lacking basic skills is problematic and must be addressed, according to a member of the higher-education group, "There needs to be some kind of foundation laid at the beginning when they first walk into the classroom. In terms of identifying, you need to start before third grade. You need to start in kindergarten." The business/community group expressed similar concerns and shared knowledge of programs from schools that are doing this effectively.

> I know some school systems, who sit down with the kids individually, and they are divided up by quarters—they have to meet your standards that are written out, they have to count from 0 to 30, by first quarter. The kindergarten teachers sit down with that kid with a print-out and one-on-one, test every kid in that class. Yes, it is time consuming for that individual teacher but they can then identify the student's weakness and use it as a positive tool and things to work on—and the notes are sent home.

Interventions should take place, not only in elementary school, but also throughout a student's academic career. Members of the teacher and administrator groups discussed this in detail and the need for differentiation to meet the needs of all. A teacher-group participant shared that "engaging every student can be done through differentiation."

Better STEM Teacher Preparation

Teacher preparation was also a topic of discussion and was identified as a need for the groups. Unfortunately, many existing teacher-preparation programs do not provide immersion in best practice, and many focus groups discussed that just the opposite may happen in many cases, as students are exposed to lecture after lecture at the university level. One member of the higher-education group gave an excellent example of the collective group discussion and the difficulty of a one-year student teaching experience and science and/or mathematics teaching methods courses to undo a lifetime of bad examples.

> When they walk into a science classroom, they only encounter lectures, we all know that there are "x" number of our professors who are lecturers. And that is one effective method, but it doesn't reach all the students. If that is what they got all the way through high school, and when they come through four years of college to somebody saying in one or two classes we are going to change that because you have gotten them used it. And when they go out with these new ideas and these new activities and they are faced with their fifty-minute period and everybody in the hallway teaching in the easy way, just stand and deliver.

Other concerns for the groups included the lack of substance of content in the preparation of elementary teachers. A member of the business/community group shared the following concern, "Maybe I don't understand the teaching profession, but I think you don't have to know much math and science to teach third grade. I think if we lobby for them to have a broader education that would help." Members of the administrator and K-12 teacher groups echoed these concerns, as one teacher shared, "If we are going to prepare the teachers to teach differently, to do things differently, then we need to think about how we educate people in college."

Increased Exploration

The manner in which STEM is taught in K-12 schools was a big concern for participants in all of the focus groups. There was an overwhelming desire to move away from memorization of isolated facts and toward opportunities for students to actually do more STEM, particularly science. One business/community group shared that K-12 schools should, "Stop teaching facts. Start teaching experience." Another member of the same group added, "I think more than that really, to give students an opportunity to do science because I don't think they do science much."

There was agreement across the groups that more extended time for investigations and real-world exploration of problems and engagement in projects should be the focus of teaching STEM disciplines. Inquiry was presented as the most effective instructional strategy that was lacking in schools. It was proposed by a K-12 teacher-group member that there is a need to "start doing more inquiry-based, problem-based teaching and learning and I would stop pure rote memorization and teaching to the test."

Professional Development

One concern was the need to provide high-quality professional development experiences to enable in-service teachers to get up to speed with the latest strategies for teaching science and providing STEM experiences for students including problem-based, real-world experiences with technology and other learning tools. One way to engage teachers in professional development was through professional learning communities, as one administrator shared, "The other thing too though, as educators, we need to start doing more professional learning communities rather than trying to teach in isolation." Support was also argued as a need for teachers to transform their classrooms into more inquiry-based environments. As one business/community member shared, "Not every person who stands up in the classroom has the skills or the confidence to engage a class of thirty that way. When they do that, it doesn't just happen. If you have never taught it before and you don't understand the method, then how can you deliver this to students?" Providing support for veteran elementary and middle school teachers was expressed as a need, as they received their preparation many years ago, and it may not have included a science or mathematics methods course work. One K-12 teacher-group member explained,

> And that goes back to those teachers that probably have been certified in the last eight to ten years that are getting more of the instruction of the hands-on subject specific, as opposed to the teachers from fifteen to twenty years ago where they were K-8 certified and their specialization may not have been math, but reading, social studies or English, but they're forced to teach a math and science.

A final need for professional development is the desire to teach the K-12 teachers and administrators how to use high-stakes assessment and diagnostic-test data effectively to improve instruction on an individual level for students. A K-12 administrator shared, "We need a little help with the data, and driving our teaching by utilizing the data that is basically available but again, as teachers to take the time to dig it up and apply. We need some help."

Strengthening STEM Partnerships

Funding, Resources, and Involvement

Partnerships can bring a lot of resources to the table, whether they are in-kind contributions of expertise or utilization of existing resources. The focus groups discussed how to leverage these resources, as well as how to leverage partnerships to apply for and secure competitive funding from state and federal agencies. One of the K-12 teacher participants shared her experiences with obtaining in-kind donations from business/community partners.

> I think there are a lot of businesses out there that replace their computers every couple of years and they would be happy, I think that there are a lot of businesses that are looking for places, and if we would just say, "We'll take them."

Business and community organizations discussed the availability of talent that could be shared with classroom teachers to enhance their instruction and to mentor teachers to improve their content knowledge. One business member shared their investment in making a difference in the community schools, "There are a lot of people at our shop, like me, who are interested in educating people. They have to be." One great source of talent is regional higher education institutions. However, members of this group discussed the need to support involvement in schools, which is not rewarded through traditional promotion criteria.

> Those faculty, which may be the untenured, the ones that need to go up for promotion maybe the ones that are the most enthusiastic and wanting to embrace some of these concepts, but they are going to be discouraged not to. So I think there needs to be a whole paradigm shift in how we evaluate it. We can't all be evaluated by the same yardstick.

Other sources of support included preservice teachers and university STEM students who may go out and work with teachers. There is a need to learn how to better utilize these resources within schools so that there is mutual learning that takes place, as one higher-education group member shared.

> That is where it is worth getting experts out. That is worth providing, if it is methods students coming out, bringing a unit with them. And if universities, the education colleges, came with modern pedagogy the teachers, as you say, will follow.

Resource allocation and sustainability of efforts were also identified as needs for partnerships. A K-12 administrator shared enthusiasm for a program that faded when the funding ended and the program was not sustained.

> They did a fabulous job making a link between university and the school systems. I don't know if there is anything like it. When the funding went away, then the program went away, so maybe there is something we could do seek funding for that kind of opportunity.

Parental Involvement

Parents are an essential part of the partnership equation. This was clear from the conversations of all of the focus groups as one business/community participant shared, "We can't fix the school system without parent buy-in. That is a requirement." The first need was to have the parents be on the same page with other stakeholders regarding the importance of STEM skills to the future of their children.

> I don't know what we are doing with our students. They want to be spoon-fed. They want to know what it is that they can do to pass the test, which is this whole state testing. So again, we have to look! If we don't look at what the parents do and educate the parents on how important this is—it doesn't matter what we do! It starts at home. It starts with their commitment to knowing that the occupations and the sciences and math are so important. And it is the language, look at Intel and all of that. (K-12 teacher group)

One of the administrators from the focus group shared the sentiments of her group including the realization that parents are the key to success in schools.

> You have to understand that yes, they spend a lot of time with us in the school. If they are not getting that positive reinforcement or that feedback at home, you can continue to talk to people until you are purple in the face and give them all the praise that you can and some of them are not going to get it.

Real Partnerships—Not Superficial

The discussion of all stakeholder groups indicated both a need and desire to strengthen partnerships and relationships across the stakeholder groups. The notion of "working smarter, not harder" to make a difference in the educational system and infuse not only the teaching of science, but also STEM experiences for students based in the real-world context. Interestingly, the four focus groups seemed confident that through building real, sustained partnerships, that the change they desired could be accomplished. One of the participants from the higher-education group articulated, "I think that developing that systemic approach, taking all these ideas where we truly have a wide approach with a lot of people involved to make it work." One business-group member shared an example of such a partnership that is in place in Michigan that has accomplished this systemic change and the way students are prepared for STEM careers.

> Ford Motors runs a charter school on the grounds at Greenfield Village, in the Henry Ford Museum. The folks from the plant—right across from the parking lot are in and out of there all day long. They are preparing their entire next workforce.

The groups believed that true partnership included all partners being co-constructors of the curriculum and moving beyond simply providing funding or resources to schools, as one K-12 educator shared, "If we believe this is the way to go, we would become partners in the construction of these lesson plans. Then we can impact what are in those lesson plans."

> Flexibility in partnerships was also discussed as a need. A good example of this flexibility is university partnerships that cater programming to teacher needs. One K-12 administrator shared an example, "If you want practicing teachers to enhance their content knowledge. Then science departments will have to offer science courses after 5:00 at night."

STEM Recruitment, Placement, and Retention

Early and Ongoing Career Awareness

In the region this study was conducted, there was a great deal of what many stakeholders referred to as "brain drain," where the more promising students who graduate from local universities, often leave the area for more attractive employment. Automobile manufacturing had been the foundation of the local economy that had seen better times and many people living in the region were losing their jobs as technological innovations transformed the way work was completed. Stakeholders determined that recruitment, placement, and retention of STEM talent, including STEM teachers, were large concerns at this time. The quantity of graduates has not been sufficient to fill vacant engineering and teaching positions to name just a few. Stakeholders in the focus groups believed that there is a need to begin in elementary school with introduction to STEM careers as a potential solution to improving the pipeline. One of the K-12 teacher participants shared his own personal experiences relating to this need.

> Most of these kids, when I survey them or ask the question at the beginning of class, "What would you like to be?" I hear, "I want to go into the medical field," "I want to be a doctor," "I want to be a heart surgeon," I want to be a lawyer" or things like that. As I bring in volunteers to share with the class about their careers, we just had the last one last week it was a health science career recruiter. And I said, "Look, we have a lot of you who want to go into the medical field and you are saying you want to be a doctor. This person is going to talk about medical fields that you don't hear a lot about in the media. Things like physical therapists, exercise science, physiology." He really opened their eyes to new areas that they just don't hear about a lot.

Members of the business/community group shared that it is simply not enough to just read about careers through texts and articles. Students need opportunities to interact with people in STEM careers so that they can identify with role models and learn more about the day-to-day of each potential career option. The other groups also believed this is a key component of an early career program. An engineer in the business/community group shared his perspectives on this topic.

> In terms of motivating the students, not necessarily preparing them for a potential career, most of them have absolutely no clue of what a life in one of these professions is like, and the material alone is not going to grab them. They want to know what it would be like if they devoted their life to one of these professions. They know what it would be like to be a doctor. They have no clue what it is to be a research scientist or an engineer, in most cases. So some people could come in and talk to them about their life, not just what they do, but how they live and how it all ties together. Particularly younger ones with which students are more likely to identify. I think this is really important.

A member of the higher-education group shared, "I think, for example, a lot of these students are interested in pharmacy because they see pharmacists, but the trouble is they don't see a physicist. What does a physicist do?"

Intern Experiences and Role Models

Meaningful introductions to the STEM careers and engaging with role models should take place both within the classroom and in alternative learning environments, including internship experiences. The participants in our focus groups identified that there was a growing need for meaningful internships and purposeful pairing of children with role models within the region. One of the business/community members shared their work in the area that has attempted to provide these types of experiences for children.

> What we are trying to do right now, we are attacking this problem humanly within our human supply chain. What we have done perfectly with success thus far is with the university we have traveled up stream and gotten much more proactive and aggressive in utilization of a co-op and intern program as a strategic part of our work force. It has had all kinds of benefits for our firm and I think for the university in the region in doing that.

The importance of having STEM role models was discussed in all of the focus groups as an integral part of shaping student interest and persistence in STEM. One of the K-12 teachers in our focus group shared, "You know your kids, and I know my kids, they don't have the role models. They don't have the relatives, the neighbors. They don't have the people in their lives that work in high tech jobs that have the background that could expose them to it." The K-12 administrator group had an extensive discussion about how to establish a mentoring system for the region. A member of that group argued for making this a critical piece of the regional work.

> I think one of our groups should focus on building confidence in these young adults, because they can do anything, if they believe in themselves. They come in saying they believe in themselves, but the confidence is not there. Somehow, someone has to find a way that little tiny steps can be made to build their confidence and I think we can do a lot.

Early STEM Learning and Experiences

There was much discussion throughout the four focus-group sessions regarding the lack of STEM experiences for children in the regional K-12 schools. Notably, this is more prevalent in the earlier years (K-6), and the need for integrating more, high-quality opportunities for children was seen as critical. Some discussion took place regarding what other countries are doing with respect to engaging children early. A member of the business/community group shared an experience in Japan. "One visited a plant in Japan, and was surprised to see, about one hundred third graders in the plant being shepherded through, learning what manufacturing was all about and what they needed to go forward." The quality of STEM learning experiences was also discussed by the various groups. One concern often raised was the need for substantial experiences that involve students doing STEM. This was articulated well by a member of the higher-education group.

> We are looking at trying to get in there and develop something where they can start working on CAD early and doing real work. That is the other key to this thing, it is not enough to hire them and then have them come in and act as clerks and gophers. That is not doing much other than putting some money in their pockets. We would hope to get them involved in doing real project work and learning in the process of doing that.

It was also discussed in the higher-education group, as well as other groups, how essential it is for children to have extensive STEM learning experiences in K-12 to prepare them to be successful in postsecondary study.

> I think that if students were learning the science at the high school level, and I know it is what came first the chicken or the egg, but if we can get

> them to get the science in the high school—fun, hands-on, problem based, inquiry-based, everything else... when they come into the chemistry classes at the undergraduate level, and it is 200 students where it is going to be more lecture. Their knowledge base is going to be a lot greater to start with and they will be able to handle a big lecture and then they could make good use some of their recitation courses or their smaller lab times because that is where the learning will occur.

One of the most important reasons for early STEM experiences argued by the various focus groups was to generate an interest in STEM disciplines. A K-12 teacher in the focus group shared her frustrations regarding lack of student interest in science.

> I asked my students, "How many of you like science?" They don't raise their hand, yet every small child asks, "Why, why, why?" and that is science. Or "How" and that is science. And yet by the time they get to me, somehow, they have lost that and they have spent the majority of their time with us. I wonder what we do.

A final justification for early childhood STEM experiences was the argument made by many of the groups that early experiences may alleviate science and STEM learning anxiety. As on K-12 administrator shared, "The kids would like it and it would decrease anxiety. They have science anxiety, just like they have math anxiety. If they are exposed to it from an earlier age, at a much higher level, then the anxiety, I think, would decrease."

Strategies

The four focus groups were followed by a regional STEM event where each participant had the opportunity to sign up for a breakout group and continue the discussion around one of the five identified needs of the region: Accountability for STEM Learning, STEM Awareness and Advocacy, Improving STEM Instruction, Strengthening STEM Partnerships, and STEM Recruitment, Placement, and Retention. The resulting heterogeneous groups included participants from K-12 teachers, K-12 administrators, business/community members, and higher-education members who were involved in breakout group work for three hours. At this event, each group was presented with the subcategories that were identified in the focus groups. They were challenged to develop an action plan that would detail potential strategies that could be used in the next six to twelve months to begin to address some of the needs. Following the event, there were monthly meetings of each group to continue the discussion and work. In the next section, the strategies that emerged from the group discussions, as well as planned deliverables for the first 12 months of work will be shared.

Accountability for STEM Learning

The Accountability for Learning group was charged with addressing STEM reform needs in the areas of accountability (holding schools and teachers responsible) for effective STEM teaching (e.g., engaging students in inquiry, problem solving, asking questions, exploration), the teaching of science, using data-driven instruction, alignment between high school and university, and instruction not driven by assessments. The first strategy that emerged from this group was to communicate the importance of effectively teaching science K-12 on a daily basis. A second strategy was planned to develop a support system for teachers and parents to use data to increase student learning in STEM. A third strategy was to assemble an alignment group that would work to align high school mathematics course work with entry-level course work at the local university. A fourth strategy was to begin a conversation with key stakeholders about the impact of high-stakes assessments on the teaching and learning of STEM content.

The group decided to focus on the strategy of "communicating the importance of teaching science K-12" as their first action item. The need for developing a regional communication tool and beginning to work on a STEM booklet was determined. This booklet would profile the importance of science learning experiences, highlight local STEM careers, discuss the STEM skills needed for success in K-12 and higher education, identify local STEM resources, and discuss potential regional STEM careers. The nature of the work of this group overlapped with the work of the STEM Awareness and Advocacy group, as well as the STEM Partnership and STEM Recruitment, Placement, and Retention groups, as their deliverable was a communication tool that met many needs concurrently.

STEM Awareness and Advocacy

The tasks of the STEM Awareness and Advocacy group included meeting the needs raised in the areas of parental and community STEM knowledge and engagement, STEM as an economic development priority, and the marketing of STEM. Strategy one identified by this group was the need to develop a framework to guide parental engagement. The local PK-16 collaborative has done some work in this area, and the STEM initiative would leverage this resource and assist the group in a STEM specific version.

Strategy two was to identify readily available resources in the community and facilitate access to parents, teachers, and schools. Strategy three was to develop and implement a STEM program for elementary schools.

The STEM Awareness and Advocacy group decided to address strategies one and two, in year one. A partnership with the PK-16 group to discuss development of the parental engagement framework began. This group collaborated with the Accountability group to include STEM resources within the STEM booklet.

Improving STEM Instruction

The Improving STEM Instruction group was assigned seven areas of need to address in their work that included the following: funding resources and support, innovativeness, rewarding excellence, improving interventions, improving STEM teacher preparation, increasing exploration, and professional development. The first strategy developed was to leverage existing resources and identify new opportunities through working with partners. Strategy two was to provide teachers with real-world opportunities to do STEM research in the field. The third strategy was to provide high-quality, engaging STEM learning experiences for students through partnerships with higher education and the community. Strategy four was to incentivize high-quality STEM teaching through merit pay. The fifth strategy was to utilize data to design individualized instruction and to bridge achievement gaps. The sixth strategy was to reform the teaching of course work for future STEM teachers to incorporate modeling of more student-centered strategies. The seventh strategy was to provide professional development experiences for teachers to support the utilization of STEM pedagogical content knowledge.

The group chose to focus their efforts on strategies one, three, and seven, during the first year of work. Partnerships with business organizations and funded programs through the state provided student programs such as science fairs, inquiry professional development for teachers, and school resources for doing STEM. Funding was secured through state grants to deliver a STEM program for students through using technology and partnering with the local university to teach STEM to students virtually.

Strengthening STEM Partnerships

The charge of the Strengthening STEM Partnerships group was to address needs in the areas of funding, resources, and involvement, parental involvement, and forming real partnerships. Strategies that were developed included: (1) engaging business/community/university partners in sharing their expertise with schools, (2) identifying business/community/university existing resources that could be leveraged for maximum impact, and (3) establishing reciprocal, meaningful, and lasting partnerships. In the first year of work, the Strengthening STEM Partnerships group decided to work on building a regional STEM collaborative partnership that would be the foundation of the work to come.

STEM Recruitment, Placement, and Retention

The STEM Recruitment, Placement, and Retention group was focused on needs relating to early and ongoing career awareness, internships and role models, and early STEM learning experiences. The first strategy that the group proposed was to develop STEM career-awareness literature that could be used with parents and community members. Strategy two was to develop elementary STEM career programs in partnership with STEM career providers. Strategy three was to institute formal internships and mentoring programs with higher education and business/community STEM stakeholders. The fourth strategy was to develop and implement early STEM learning experiences in and outside of schools.

The group chose to focus their first-year efforts on strategy one and four. Strategy one was rolled out in partnership with participants in other groups through the STEM booklet that was developed for the region. A specific focus on profiling people in STEM careers, as well as projections of STEM career needs were included in the booklet design. Strategy four was implemented in collaboration with the Improving STEM Instruction group through their student STEM program that was planned and delivered by a partnership between local schools and higher education.

Discussion and Implications

STEM stakeholders in this study approached the formidable challenge of reinventing the educational system in northwest Ohio with eyes and minds wide open. Perhaps one of the work-group members described the group consensus best when she stated, "I think that developing that systemic approach—taking all these ideas where we truly have a wide approach with a lot of people involved to make it work." Participants in this study were ready and willing to take a critical look at the way the business of education had been done in this region for many years. For the most part, there had been a hands-off approach of working with the urban K-12 school districts in the region and little motivation to build partnerships other than in isolated pockets. It is important to note that this was a regional effort. However, there were urban, rural, and suburban, as well as public and private schools that were included in the effort—thus creating an array of backgrounds and levels of support and excellence—though many commonalities were discovered across the groups.

Lewin's (1947) change theory was a good lens to observe the process in this region. Participants in the focus groups enacted the first step of unfreezing the status quo by getting the issues on the table and discussing the needs that must be addressed to move the region's STEM educational system forward. Interestingly, the five primary areas of need were not new revelations for the research community. These concerns have been prevalent in recent discussions of STEM reform. The persistence of these issues signals that although we are very aware of their existence, we have yet to formulate a response that has been disseminated and implemented on a larger basis to promote reform. Accountability for learning was one of the areas of need that was identified by the constituent groups, which included both STEM teacher quality and the increasing challenges of reduced science instructional time within elementary classrooms. There has been extensive literature supporting the influence of and the need for more effective STEM teachers (i.e., Committee on Prospering in the Global Economy, 2007; Johnson, Kahle, and Fargo, 2007b), as well as the detrimental impact of students failing to receive a strong foundation in STEM learning experiences (i.e., Business Roundtable, 2005; Carnegie Foundation, 2009). Additionally, the failure to align high school course work with university entry-level course work deters students from entering STEM fields of study (i.e., Business Roundtable, 2005; CSEPP, 2007).

At the policy level, STEM is an economic development agenda item of high priority, which has begun to trickle down into the education agenda as well (U.S. Commission on National Security/21st Century, 2001; Obama-Biden Plan, 2009). This was one of the focus areas in the needs category involving STEM Awareness and Advocacy. This is a reality in which stakeholders in this region are far too aware. What they call the "brain drain" is unacceptable, and the need to repair and enhance their regional STEM talent pipeline was articulated. As one member of the business community shared, "You need to put together an educational system that puts out some critical thinkers that know science, math, and technology along with language arts skills. I think if you look at the number of jobs that they are projecting people are going into—we need to do that." Parental and community engagement were needs that overlapped more than one category, but were a primary target group within the awareness-need area, particularly for urban school districts in the region. A K-12 administrator shared the need for educating parents on how to support their children in STEM, "Parents can't really help their students in the way that they need to. So, parental knowledge is critical."

In addition to improving accountability for STEM teaching in the region, there was also great emphasis placed on the primary-need area of Improving STEM Instruction. Within this area of need, many things that have been prevalent in the literature on improving STEM in schools. Several thoughts emerged including funding, resources and support for both teachers in urban and rural, as well as suburban districts in the region (Ruby, 2006; Johnson, 2007b; Johnson, Kahle, and Fargo, 2007a). The need for teachers to be innovative and teach for the jobs of the future, including multiple experiences for students to engage in exploration, was also key to moving the regional STEM educational system forward (i.e., Business Roundtable, 2005; Wallis, 2006). Interestingly, there was extensive support for rewarding STEM teacher excellence through merit pay and other perks to encourage teachers to go above and beyond. Need for innovative STEM teacher preparation at regional institutions of higher education and professional development was also discussed (Committee on Prospering the Global Economy of the 21st Century, 2007; Carnegie Foundation, 2009), as was the ever-present disconnect between much of the schooling experience for preservice and inservice teachers (including university course work) and the desired innovative, student-centered practice (Wright and Wright, 2000).

Partnerships for STEM were an area of need whose subcategories overlapped with other areas. Funding, resources, and the engagement of partners were needs that were discussed as critical to sustaining efforts of the group. As one K-12 teacher participant shared, teachers value when partners are able to, "bring the facilities to the students through mobile outreach." Engaging parents as part of the partnership was also seen as essential. Establishing real partnerships that go beyond a project-to-project basis was seen as the most critical component.

STEM Recruitment, Placement, and Retention were areas of need that were identified by the participants in this study. As the Committee on Prospering in the Global Economy of the 21st Century (2007) argued in *Rising Above the Gathering Storm,* recruitment of students into the STEM pipeline should begin early. In this study, early and ongoing career awareness and STEM learning was identified as pivotal to moving forward (CSEPP, 2007; Carnegie Foundation, 2009). A member of the higher-education focus group reinforced this need as he discussed, "I mean this is what kids need.... They are so excited. They love science when they are in those primary years." As students move through K-12 schools, intern experiences and STEM role models become increasingly important (Obama-Biden Plan, 2009).

The STEM stakeholders in this study progressed to Lewin's (1947) second step of change as they moved into the regional STEM event and the accompanying monthly follow-up meetings where they crafted their strategies for addressing the collective needs and began construction of the deliverables. In this period of transition, the participants in this study are moving from the status quo that has been unfrozen, and they are working together to determine what the new system will entail. Partnerships and leadership to support the desired change also began to emerge from the various groups.

Although the STEM stakeholders in this study were positioned to challenge and unfreeze the status quo and move toward developing what the new STEM educational system will be in the future for this region, the third step of refreezing did not occur during this one-year study. Arguably, this may take more time to be crafted with more input from the various constituent organizations over time. It is assured that the motivation for change is high and progress will continue until the collective group comes to consensus and begins to weave the innovations into the fabric of the system.

In the bigger picture of contributions to advancing a theory of change for science education reform, as Fullan (2006) shared, "We have been using and refining our change knowledge over the past decade, in particular in order to design strategies that get results" (p. 6). Fullan (2006) argued that there are seven core premises that comprise our use of change knowledge that were shared in our theoretical frame for this study. We found motivation, capacity building, and reflective practice to be the most crucial in this study of a regional reform effort. Fullan (2006) included tri-level engagement as one of the core premises. However, we found in regional reform movements, state support is not crucial in science education reform initiatives. This may be due to disconnect between the standards, accountability, particularly AYP, and an emphasis at the state level on science. Additionally, Fullan's (2006) work appears to have only a secondary focus on community, whereas we found the community engagement component to be a critical and primary requirement. This can be attributed to the region in particular, similar to many other areas across the country, whose engagement and motivation are driven by the local economy and a shortage of talent with STEM skills, especially science backgrounds. This region has collectively realized the impact the lack of quality science experiences has on the critical thinking and innovativeness of the future generation.

Many, who think of STEM, think of science education. This may be due to the lack of emphasis that the discipline of science has received over the past several years due to the increased emphasis on mathematics for AYP. This regional study echoed national concerns and warranted focus. Many schools and curriculums across the nation are now increasingly STEM focused. Science is primarily the discipline where integration of STEM disciplines occurs. Arguably, science is the anchor for STEM, and the weakened STEM talent pipeline in the United States is promulgated by the quality and quantity of science experiences our children receive in our schools today. Therefore, there is a strong argument that science educators should pay attention to STEM and leverage the STEM movement as an opportunity to infuse science as the focal point.

Some lessons learned from the work in this study may help to inform other regional STEM reform efforts. These include: (1) large-scale regional grass-roots discussions may result in more movement within the educational system than decades of top-down initiatives that fade quickly before institutionalized, (2) there must be buy in and leadership, not only from school systems, but also communities and other supports such as higher education, to truly transform the educational system and the STEM talent pipeline, and (3) progress, even slow progress, can be beneficial as stakeholders buy in to change and this begins to motivate their conversations and commitment.

Postscript: Targeting Turbulence: Lessons Learned—Potential Solutions to Challenges

Carla C. Johnson

Throughout this book, we have highlighted some of the ways that *turbulence* can influence STEM project implementation and success. In the introduction, we described the concept of turbulence as it relates to educational reform, as well as provided examples of the types of turbulence and some of the briefly discussed challenges that our authors had encountered in their programs due to associated turbulence. As we close this book with this last section, we will present some suggested solutions to challenges created by turbulence within secondary STEM educational reform.

Micro-level turbulence can include personnel, scheduling, student support, learning environment, accountability, and community engagement. Primary sources of macro-level turbulence are accountability and funding. Authors of the chapters in this book promulgated the necessity to deal with turbulence and provided examples of potential solutions that can be enacted to better support reform efforts.

Personnel issues such as workload, salary, job assignment, and opportunities for professional growth are often producers of turbulence. In chapter 2, the authors proposed providing time within the school day for delivery of professional development that is aligned with the reform program. In chapter 4, creating teacher leaders was a potential strategy to scale-up district-level ability to deliver ongoing professional growth opportunities. Yore et al. (2007) also described the need to provide time to teachers within their workloads for planning and less preparations to encourage teachers to refine their practice. In chapter 8, leveraging community/business partners was suggested as a solution to challenges of resources for professional development. Additionally, the use of incentives such as merit pay to encourage high-quality teaching and retain talented staff were proposed.

Scheduling has also been revealed as a source of turbulence primarily in the areas of class period duration, time for meeting with colleagues, and opportunities for planning and collaboration. In chapter 4, establishing communities of practice within and across schools is a potential solution to challenges of lack of time for planning and collaboration. These communities operate informally and are often great supports for teachers who cannot find the time within the school day. Chapter 6 also pointed to the building of partnerships and co-teaching through breaking down silos within and across institutions. Often scheduling decisions are made based upon budgetary rationales rather than deciding what is best to support high-quality teaching and student learning. Chapter 8 calls for generating advocacy for the importance of STEM and use data to support student learning—particularly in science that has been minimized for many years at the elementary level. Encouraging district and school administration to repurpose staff meetings to focus on data and strategic planning rather than administrative tasks and delivery of information is another strategy to provide more time for teachers to collaborate.

Turbulence can also be created through a lacking student support policy and policy implementation. Our authors provided several solutions to student support challenges. In chapter 3, the development of a sharing and delivery system for the purpose of sharing enrichment, remediation, English as a Second Language (ESL), and special education strategies across schools from more-resource rich to less-resourced schools. Chapter 4 proposed schools partner with community organizations and higher education to provide enrichment opportunities for students. Chapter 6 argued that STEM teacher-preparation programs should have a more purposeful focus on developing literacy to address needs of all students. Class size is more elusive for external partners to control. However, one suggestion we have is to use data to demonstrate the difference in student learning between large and average class sizes to persuade leadership to rethink this policy.

Resources and facilities are the two forms of turbulence within the learning environment domain. Chapter 1 proposed ensuring funded programs purchase new curriculum materials for use in reform schools and engaging community partners in providing resources for improving facilities. Chapter 8 argued leveraging partnerships to deliver high quality STEM real-world learning experiences for students when resources are short in schools.

Accountability is both a micro and macro source of turbulence. At the micro level, districts have responded to the need to make gains on state assessments by developing pacing guides to keep all students on the same page and district assessments to measure progress at multiple points across the academic year. Chapter 3 proposed using the data from district assessments to drive areas of focus for teacher professional development. Chapter 8 suggested using the data to construct and deliver individualized instruction. Pacing guides can hinder the ability to individualize instruction—as all students are expected to move forward based upon the schedule. An additional solution to utilize pacing guides is to begin the year or term with a baseline measure of what students know—then using this prior knowledge base as the road map for planning future instruction.

Lack of parental and/or community engagement was echoed across the chapters as a source of turbulence as well as a potential key solution to improving STEM education. Chapter 2 argued that partners are integral in the reform process and used partners to help with technology integration, school beautification, and the decision-making process. In chapter 5, the authors proposed making parents a partner in enrichment programs to facilitate student engagement and retention. Bringing partners to the table and making them partners in the administration of the reform project is a solution Chapter 7 proposed to have clear communication of goals, roles, and task completion. In chapter 8, a solution to engagement turbulence proposed was to develop and disseminate a community/parental engagement framework to support new partners come on board.

Macro-level policy is often outside of the reach of STEM reform programs. However, there is power in numbers and as chapter 8 suggested, building a clear STEM advocacy program where stakeholders across regions, states, and the United States are communicating the importance of STEM reform and securing the resources needed to make reform a reality may be a solution to this challenge. Accountability in principle is not bad. However, accountability measures are driven more by the bottom line rather than quality of knowledge measured. Thus, the reason why funding is the other source of macro-level turbulence. In order to stop the cycle of teaching for the test and memorization of isolated facts for the purpose of gaining a grade or score needed for graduation, et cetera, we must reform our accountability system and invest in high-quality applications of knowledge rather than recall.

This book, *Reforming Secondary STEM Education* has many implications for practice. We have shared challenges across a diverse set of programs that created turbulence for stakeholders within each project. Potential solutions to address turbulence were argued. It is our hope that the reader will walk away with a sense that STEM reform is not easy, and there are factors that must be addressed that are outside of the realm of the project staff. This book is the first step—talking about the elephant in the room—and beginning to develop a plan to navigate the turbulence and avoiding bumps in the road that could derail reform.

Appendix 6.1

Interview Framework

Each type of interview used in data collection is briefly described below. Sample open-ended questions are provided for each type interview described.

Executive Board Members' Interview Protocol

The open-ended, semi-structured interviews lasted 45–75 minutes. During the interviews, the researcher focused on the research participants' background and vision and perspective of the Executive Board, along with perceptions of STEM issues in the city.

1. What can you tell me about yourself?
2. Why does Science, Technology, Engineering, and Mathematics (STEM) education matter?
3. What are the key issues of STEM education in the metropolitan area?
   a. What concerns you most about the current status of STEM teacher education in the region?
4. What role do you believe that public and business have in increasing the number of STEM graduates?
5. What can you or your company/institution do or provide to ensure that K-16 schools better prepare students for STEM jobs in the region?
6. What can you tell me about the purpose of executive board? As a board, what are you hoping to accomplish?

7. What roles have you taken as a member of Executive Board?
   a.What did make you decide to join the board?
8. In your opinion, what key group of people can be counted on to assist in conveying the Executive Board's message? Why?
9. Based on the experiences you have had so far as a board member, what is necessary to have a functioning Executive Board aiming to initiate and implement improvements in K-16 STEM talent development and STEM teacher education in MTA?
   a. What do you believe that have challenged or constrained Executive Board in doing so?

Executive Board Director's Interview Protocol

The open-ended, semi-structured interview lasted 60 minutes. During the interview, the researcher focused on the director's background and vision and perspective of the Executive Board, along with perceptions of STEM issues in the city as well as her expectations of the board members, along with the challenges to effectiveness of the board. This sample illustrates some of the interview questions that are unique to the director.

1. What is the purpose of the Executive Board? What do you want to accomplish as the director?
2. What difficulties and constraints have you faced as you put together the Executive Board together?
3. What kind of roles do science and mathematics teacher education programs have in increasing the number of STEM graduates? What role(s) can the Executive Board play?
4. Based on the experiences you had so far, what is necessary to have a functioning Executive Board in higher education or educational settings aimed at improving STEM education?
5. What roles you have taken on the Executive Board besides initiating it?
6. What politics are involved in this process? How have you handled them?
7. To your knowledge, how are the purpose and objectives of Executive Board similar or different than other works both at local and national level?
   a. What is new?
   b. What is different?

8. What advice would you have for someone who wants to initiate the same thing elsewhere?
9. Have there been any changes in the intended objectives and goals since you began?
   a. What are these changes?
   b. What prompted these changes?

Appendix 7.1

Focus-Group Protocol

K-12 Teachers and Administrators

1. What do we want to *keep* doing, *start* doing, and *stop* doing as far as current practices in schools to educate students in STEM?
2. What do we want to *keep* doing, *start* doing, and *stop* doing as far as support for professional growth for teachers and administrators to enhance their instructional skills and ability to develop STEM talent?
3. What do we want to *keep* doing, *start* doing, and *stop* doing related to classroom experiences which enable all students to succeed and gain a strong background that prepares them for college and STEM careers?
4. How can we work to solve the STEM talent problem that we are experiencing in our region, the state, and the nation?

Higher Education

1. What do we want to *keep* doing, *start* doing, and *stop* doing to market STEM careers and provide experiences for students to learn what these careers require?
2. What do we want to *keep* doing, *start* doing, and *stop* doing related to university faculty support for PK-12 classroom teachers?
3. What do we want to *keep* doing, *start* doing, and *stop* doing with incentives at the university level for improving teaching in their classes?
4. What do we want to *keep* doing, *start* doing, and *stop* doing related to experiences to prepare future teachers for the PK-12 classroom?

5. How can we work to solve the STEM talent problem that we are experiencing in our region, the state, and the nation?

Business/Community

1. How can we work to generate awareness in our region of the need to better prepare our children with the skills necessary to compete in the global society?
2. How can parents and families better support their children to pursue STEM careers?
3. What is the best way to bring the community up to speed about the requirements to be successful in the new knowledge-based economy?
4. What are our biggest problems associated with job needs in our region?
5. What are the skills that the workforce will need in the future to be successful?
6. How can we work to solve the STEM talent problem that we are experiencing in our region, the state, and the nation?

Bibliography

Achieve, Inc. (2007). Creating a world-class education system in Ohio. Unpublished report, Columbus, OH: Ohio Department of Education.

Adobor, H. (2004). High performance management of shared-managed joint venture teams: Contextual and socio-dynamic factors. *Team Performance Management,* 10(3/4), 65–76.

Aguilera, R. V. (2005). Corporate governance and director accountability: An institutional comparative perspective. *British Journal of Management,* 16(1), 39–53.

Akerson, V. L., and Hanuscin, D. L. (2007). Teaching nature of science through inquiry: Results of a 3-year professional development program. *Journal of Research in Science Teaching,* 44(5), 653–680.

American Association for the Advancement of Science. (1989). *Science for all Americans: Project 2061 report on literacy goals in science, mathematics, and technology.* Washington, DC: American Association for the Advancement of Science.

———. (1990). *Science for all Americans: Project 2061.* New York, NY: Oxford University Press.

———. (1993). *Benchmarks for science literacy: Project 2061.* New York, NY: Oxford University Press.

———. (1997). *Blueprints for reform: Project 2061.* Retrieved from http://www.project2061.org/publications/bfr/online/blpintro.htm.

Anderson, K. M. (2007). Differentiating instruction to include all students. *Preventing School Failure,* 51(3), 49–54.

Anderson, R. D. (2002). Reforming science teaching: What research says about inquiry. *Journal of Science Teacher Education,* 13(1), 1–12.

Annetta, L. A., Folta, E., and Klesath, M. (2010). *V-Learning: Distance education in the 21st century through 3D virtual learning environments.* Dordrecht, The Netherlands: Springer.

Annetta, L. A., and Shymansky, J. A. (2006). Investigating science learning for rural elementary school teachers in a professional-development project through three distance-education strategies. *Journal of Research in Science Teaching,* 43(10), 1019–1039.

———. (2008). A comparison of rural elementary school teacher attitudes toward three modes of distance education for science professional development. *Journal of Science Teacher Education,* 19(3), 255–267.

Anyon, J. (2001). Inner cities, affluent suburbs, and unequal educational opportunity. In J. Banks and C. Banks (Eds.), *Multicultural education: Issues and perspectives* (4th ed., pp. 85–102). New York: John Wiley & Sons, Inc.

Attinasi, L. C., Jr. (1991). *Phenomenological interviewing in the conduct of institutional research: An argument and an illustration.* (Association for Institutional Research, Florida State University, Tallahassee, FL). (ERIC Document Reproduction Service No. ED422388).

Ball, D. L., and Cohen, (1999). Developing practice, developing practitioners: Toward a practice-based theory of professional development. In G. Sykes (Ed.), *The heart of the matter: Teaching as the learning profession.* San Francisco, CA: Jossey-Bass.

Bandura, A. (1997). *Self-efficacy: The exercise of control.* New York: Freeman.

———. (2000). Exercise of human agency through collective efficacy. *Current Issues in Psychological Science.* 9(3), 75–78.

Banilower, E. R., Boyd, S. E., Pasley, J. D., and Weiss, I. R. (2006). *Lessons from a decade of mathematics and science reform: A capstone report for the local systemic change through teacher enhancement initiative.* Chapel Hill, NC: Horizon Research Inc.

Banilower, E. R., Heck, D. J., and Weiss, I. R. (2007). Can professional development make the vision of the standards a reality? The impact of the national science foundation's local systemic change through teacher enhancement initiative. *Journal of Research in Science Teaching,* 44(3), 375–395.

Barrows, H. S. (2000). *Problem-based learning applied to medical education.* Springfield, IL: Southern Illinois University School of Medicine.

Barton, A. C. (2001). Science education in urban settings: Seeking new ways of praxis through critical ethnography. *Journal of Research in Science Teaching,* 38(8), 899–917.

Basham, J. D., Israel, M., and Maynard, K. (2010). An ecological model of STEM education: Operationalizing STEM for all. *Journal of Special Education Technology,* 25(3), 9–19.

Basham, J. D., Meyer, H., and Perry, E. (2010). The design and application of the digital backpack. *Journal of Research on Technology in Education,* 42(4), 339–359.

Bass, B. M., and Riggio, R. E. (2006). *Transformational leadership.* Mahwah, NJ: Lawrence Erlbaum Associates, Publishers.

Benbow, C. P., and Stanley, J. C. (1984). Gender and the science major. *Advances in motivation and achievement,* 2, 165–196. Greenwich, CT: JAI Press, Inc.

Birman, B. F., Desimone, L., Porter, A. C., and Garet, M. S. (2000). Designing professional development that works. *Educational Leadership,* 57(8), 28–33.

Bloom, B. J., and Sosniak, L. A. (1985). *Developing talent in young people.* New York: Ballantine Books.

Blumenfeld, P. C., Soloway, E., Marx, R. W., Krajcik, J. S., Guzdial, M., and Palincsar, A. (1991). Motivating project-based learning: Sustaining the doing, supporting the learning. *Educational Psychologist,* 26 (3 and 4), 369–398.

Braunger, J., and Lewis, J. (2006). *Building a knowledge base in reading.* Newark, DE: International Reading Association.

Brekke, J. S., Ell, K., and Palinkas, L. A. (2007). Translational science at the National Institute of Mental Health: Can social work takes its rightful place? *Research on Social Work Practice*, 17, 1–11.

Brown, J. S., Collins, A., and Duguid, P. (1989). Situated cognition and the culture of learning. *Educational Researcher*, 18, 32–42.

Brown, K. M., and Wynn, S. R. (2007). Teacher retention issues: How some principals are supporting and keeping new teachers. *Journal of School Leadership*, 17, 664–697.

Bruner, J. (1966). *Toward a theory of instruction*. Cambridge, MA: The Belknap Press of Harvard University Press.

Buck Institute. (2010). What is PBL? In *Project based learning for the 21st century*. Retrieved from http://www.bie.org/about/what_is_pbl

Bullough, R. V., and Kauchak, D. (1997). Partnerships between higher education and secondary schools: Some problems. *Journal of Education for Teaching*, 23(3), 215–233.

Bureau of Labor Statistics. (2008). *Employment projections: 2008–2018 summary*. Retrieved from http://www.bls.gov/news.release/ecopro.nr0.htm

Business Roundtable. (2005). *Tapping America's potential: The education for innovative initiative*. Retrieved July 30, 2007, from http://www.businessroundtable.org/pdf/20050727002TAPStatement.pdf

Buxton, C. (2010). Science inquiry, academic language, and civic engagement. *Democracy & Education*, 18(3), 17–22.

Carnegie Foundation. (2009). *The opportunity equation: Transforming mathematics and science education for citizenship and the global economy*. New York: Institute for Advanced Study.

Carnegie-IAS Commission. (2009). *The opportunity equation: Mobilizing for excellence and equity in mathematics and science education*. New York: Carnegie Corporation.

Cast. (2008). *Universal Design for Learning guidelines version 1.0*. Wakefield, MA: Author.

Chae, Y., Purzer, S., and Cardella, M. (2010). Core concepts for engineering literacy: The interrelationships among STEM disciplines. Paper presented at the American Society for Engineering Education, Louisville, KY.

Charan, R. (1998). *Boards at work: How corporate boards create competitive advantage*. San Francisco, CA: Jossey-Bass Inc.

Clarke, D. (1994). Encouraging professional growth and mathematics reform through collegial interaction. In C. B. Aichele and A. F. Coxford (Eds.), *Professional development for teachers of mathematics* (pp. 37–48). Reston, VA: NCTM.

Cochran-Smith, M. (2006). The new teacher education: For better or for worse? *Educational Researcher*, 34(7), 3–17.

Coleman, L. J. and Southern, T. S. (2006). Bringing the potential of underserved children to the threshold of talent development. *Gifted Child Today*, 29(1), 35–45.

Collins, A., Brown, J. S., and Newman, S. E. (1989). Cognitive apprenticeship: Teaching the crafts of reading, writing and mathematics. In L. B. Resnick (Ed.), *Knowing, learning and instruction: Essays in honor of Robert Glaser*. Hillsdale, NJ: Erlbaum.

Committee for Economic Development. (2003). Learning for the future: Changing the culture of math and science education to ensure a competitive workforce. A statement by the research and policy committee of the committee for economic development. Retrieved from http://www.ced.org/docs/report/report_scientists.pdf

Committee on Prospering in the Global Economy of the 21st Century. (2007). Rising above the gathering storm. Retrieved July 11, 2007, from the National Academies Press website: http://www.nap.edu/catalog/11463.html

Committee on Science, Engineering, and Public Policy. (2005). *Rising above the gathering storm: Energizing and employing America for a brighter economic future.* Washington, DC: The National Academies Press. Also available at http://www.nap.edu/books/0309100399/html

———. (2006). *Rising above the gathering storm: Energizing and employing America for a brighter economic future.* Washington, DC: National Academies Press.

———. (2007). *Rising above the gathering storm: Energizing and empowering America for brighter economic future.* Washington, DC: The National Academies Press.

Conger, J. A., Lawler, E., and Finegold, D. L. (2001). *Corporate boards: New strategies for adding value at the top.* San Francisco, CA: Jossey-Bass Inc.

Coulson-Thomas, C. (1991). What the personnel director can bring to the boardroom table. *Personnel Management,* 23(10), 36–39.

Council of the Great City Schools. (2003). *Restoring excellence to the District of Columbia Public Schools.* Retrieved from http://www.dcpswatch.com/dcps/0312.htm

Crawford, B. (2000). Embracing the essence of inquiry: New roles for science teachers. *Journal of Research in Science Teaching,* 37(9), 916–937.

Czerniak, C. M., Beltyukova, S., Struble, J., Haney, J. J. and Lumpe, A. T. (2005). Do you see what I see? The relationship between a professional development model and student achievement. In R. E. Yager (Ed.), *Exemplary science in grades 5–8: Standards-based success stories* (pp. 13–43). Arlington, VA: NSTA Press.

D'Ambrosio, B. D., Boone, W. J., and Harkness, S. S. (2004). Planning district wide professional development: Insights gained from teachers and students regarding mathematics teaching in a large urban district. *School Science and Mathematics,* 104(1), 5–15.

Darling-Hammond, L. (1998). Teachers and teaching: Testing policy hypotheses from a national commission report. *Educational Researcher,* 27(1), 5–17.

———. (2007). Third annual brown lecture in education research—The flat earth and education: How America's commitment to equity will determine our future. *Educational Researcher,* 36(6), 318–334.

Dede, C., Korte, S., Nelson, R., Valdez, G., and Ward, D. (2005). Transforming learning for the 21st century: An economic imperative. Retrieved August 12, 2010 from http://www.gse.harvard.edu/~dedech/Transformations.pdf

Deloitte Consulting (2005). Industry-based competitive strategies for Ohio: Managing three portfolios. Report submitted to the Ohio Department Of Development and Techsolve.

Denzin, N. K., and Lincoln, Y. S. (Eds.). (2005). *The sage handbook of qualitative research.* Thousand Oaks, CA: Sage Publications.

Dolciani, M. P. (1967). *Modern school mathematics: Structure and method.* Boston: Houghton Mifflin.

Donahue, D. (2003). Reading across the great divide: English and math teachers apprentice one another as readers and disciplinary insiders. *Journal of Adolescent & Adult Literacy,* 47(1), 24–37.

Draper, R. J. (2002). Every teacher a literacy teacher? An analysis of the literacy-related messages in secondary methods textbooks. *Journal of Literacy Research,* 34(3), 357–384.

———. (2008). Redefining content-area literacy teacher education: Finding my voice through collaboration. *Harvard Educational Review,* 78 (1), 60–83.

Draper, R. J., Smith, L. K., Hall, K. M., and Siebert, D. (2005). What's more important—literacy or content? Confronting the literacy-content dualism. *Action in Teacher Education,* 27(2), 12–21.

Education Commission of the States. (1997). *Investing in teacher professional development: A look at 16 school districts.* Denver, CO: Author.

Education Development Center, Inc. (1997). *InSights: An elementary hands-on inquiry science curriculum* (series). Dubuque, IA: Kendall Hunt.

Ericsson, K. A., and Lehmann, A. C. (1996). Expert and exceptional performance: Evidence on maximal adaptations on task constraints. *Annual Review of Psychology,* 47, 273–305.

Feldman, D. H. (1994). Creativity: Dreams, insights, and transformations. In D. H. Feldman, M.Csikszentmihalyi, and H. Gardner (Eds.), *Changing the world: A framework for the study of creativity* (pp. 85–102). Westport, CT: Praeger.

Forbes, D. P., and Milliken, F. J. (1999). Cognition and corporate governance: Understanding boards of directors as strategic-decision making groups. *Academy of Management Review,* 24, 489–505.

Ford, C. L., Yore, L. D., and Anthony, R. J. (1997). Reforms, visions, and standards: A cross-curricular view from an elementary school perspective. Paper presented at the Annual Meeting of the National Association for Research in Science Teaching, Oak Brook, IL. Retrieved from ERIC database. (ED406168).

Ford, D. Y., Grantham, T. C., and Whiting, G. W. (2008). Another look at the achievement gap. *Urban Education,* 43, 516–239.

Ford, J. D., and Ford, L. W. (2010). Stop blaming resistance to change and start using it. *Organizational Dynamics,* 39(1), 24–36.

Fullan, M. (2006). Change theory: A force for school improvement. Centre for strategic education seminar series paper no. 157. Retrieved from http://www.michaelfullan.ca/articles_2009.htm

Fullen, M. (1995). The school as a learning organization: Distant dreams. *Theory into Practice,* 34, 230–235.

Gamoran, A., Anderson, C. W., Quiroz, P. A., Secada, W. G., Williams, T., and Ashmann, S. (2003). *Transforming teaching in math and science.* New York, NY: Teachers College Press.

Gee, J. (2007). *Social linguistics and literacies: Ideology in discourses.* New York: Routledge.

Geier, R., Blumenfeld, P. C., Marx, R. W., Krajcik, J. S., Fishman, B., Soloway, E., Fishman, B., and Clay-Chambers, J. (2008). Standardized test outcomes for students engaged in inquiry-based science curricula in the context of urban reform. *Journal of Research in Science Teaching,* 45(8), 922–939.

Glesne, C. (2006). *Becoming qualitative researchers: An introduction* (3rd ed.). New York: Pearson Education, Inc.

Gold, E., Simon, E., and Brown, C. (2002). Strong neighborhoods strong schools: The indicators project on education organizing. The cross city campaign for urban school reform. Retrieved from http://crosscity.org

Gray, W. S. (1925). Summary of investigations relating to reading. *Supplementary educational monograph*, No. 28. Chicago, IL: University of Chicago Press.

Green, J. C., and Griesinger, D. W. (1996). Board performance and organizational in nonprofit social services organizations. *Nonprofit Management & Leadership*, 6, 381–402.

Grobe, W. and McCall, D. (2004). Valid uses of student testing as part of authentic and comprehensive student assessment, school reports, and school system accountability: A statement of concern from the international confederation of principals. *Educational Horizons*, 82, 131–142.

Gully, S. M., Incalcaterra, K. A., Joshi, A., and Beaubien, J. M. (2002). A meta-analysis of team-efficacy, potency, and performance: Interdependence and level of analysis as moderators of observed relationships. *Journal of Applied Psychology*, 87, 819–832.

Guskey, T. R. (1998). The age of our accountability: Evaluation must be an integral part of staff development. *Journal of Staff Development*, 19, 36–44.

———. (2000). *Evaluating professional development*. Thousand Oakes, CA: Corwin.

———. (2002). Does it make a difference? Evaluating professional development. *Educational Leadership*, 59(6), 45–51.

Hannan, M. and Freeman, J. (1984). Structural inertia and organizational change. *American Sociological Review*, 49, 149–164.

Hargreaves, A., and Fink, D. (2003). Sustaining leadership. *Phi Delta Kappan*, 84, 693–700.

Harkness, S. S., Plante, L., and Lane, C. (2008). The successes of an unsuccessful professional development program: Using the "Ohio Standards for Professional Development" to rate our efforts. *The OHIO Journal of Teacher Education*, 21(2), 29–35.

Hasselgren, B., and Beach, D. (1997). Pehnomenography—A "good-for-nothing brother" of phenomenology? Outline of an analysis. *Higher Education Research & Development*, 16, 191–202.

Haveman, H. (1992). Between a rock and a hard place: Organizational change and performance under conditions of fundamental environmental transformation. *Administrative Science Quarterly*, 37, 48–75.

Heller, R., and Greenleaf, C. (2007). *Literacy instruction in the content areas: Getting to the core of middle and high school improvement*. Washington, DC: Alliance for Excellent Education.

Henriques, L. (1997). A study to define and verify a model of interactive-constructive elementary school science teaching. Unpublished doctoral dissertation, University of Iowa, Iowa City, IA.

Herber, H. L. (1970). *Teaching reading in the content areas*. Englewood Cliffs, NJ: Prentice-Hall, Inc.

Hitt, M. A., Ireland, R. D., and Hoskisson, R. E. (2008). *Strategic management: Competitiveness and globalization, concepts and cases*. Mason, OH: South Western Educational Publishing.

Hmelo-Silver, C. E. (2004). Problem-based learning: What and how do students learn? *Educational Psychologist,* 41, 67–77.

Hough, A. (2006). In search of board effectiveness. *Nonprofit Management & Leadership,* 6, 373–377.

Houle, C. O. (1989). *Governing boards: Their nature and nurture.* San Francisco: Jossey-Bass.

Huberman, M. (1995). Networks that alter teaching: Conceptualizations, exchanges and experiments. *Teachers and Teaching: Theory and Practice,* 1(2), 193–211.

Hurd, P. D. (1958). Science literacy: Its meaning for American schools. *Educational Leadership,* 16, 13–16 and 52.

Huse, M. (2005). Accountability and creating accountability: A framework for exploring behavioral perspectives of corporate governance. *British Journal of Management,* 16(1), 65–79.

Ingley, C. B., and Van der Walt, N. T. (2003). Board configuration: Building better boards. *Corporate Governance,* 3(4), 5–17.

Innovation America Task Force. (2007). *Building a science, technology, engineering, and math agenda.* Washington, DC: National Governor's Association.

International Society for Technology in Education. (2007). *ISTE:NETS student standards 2007.* Retrieved from http://www.iste.org/standards/nets-for-students/nets-student-standards-2007.aspx

Johnson, C. C. (2006). Effective professional development and change in practice: Barriers science teachers encounter and implications for reform. *School Science and Mathematics,* 106(3), 150–172.

———. (2007a). Effective science teaching, professional development and No Child Left Behind: Barriers, dilemmas, and reality (Editorial). *Journal of Science Teacher Education,* 18(2), 133–136.

———. (2007b). Whole-school collaborative sustained professional development and science teacher change: Signs of progress. *Journal of Science Teacher Education,* 18(4), 629–661.

———. (2010). Transformative professional development for in-Service teachers: Enabling change in science teaching to better meet the needs of Hispanic ELL students. In D. W. Sunal, D. S. Sunal, M. Mantero, and E. Wright (Eds.), *Teaching science with Hispanic ELLs in K-16 classrooms* (pp. 233–252).

Johnson, C. C. and Fargo, J. D. (2010). Urban school reform through transformative professional development: Impact on teacher change and student learning of science. *Urban Education,* 45(1), 4–29.

Johnson, C. C., Fargo, J. D., and Kahle, J. B. (2010). The cumulative and residual impact of a system reform program on teacher change and student learning of science. *School Science and Mathematics,* 110, 144–159.

Johnson, C. C., Kahle, J. B., and Fargo, J. (2007a) Effective teaching results in increased science achievement for all students. *Science Education,* 91(3), 371–383.

———. (2007b) A study of sustained, whole-school, professional development on student achievement in science. *Journal of Research in Science Teaching,* 44(6), 775–786.

Johnson, C. C. and Marx, S. (2009) Transformative professional development: A model for urban science education reform. *Journal of Science Teacher Education.* 20(2), 113–134.

Johnson, H., and Freedman, L. (2005). *Content area literature circles: Using discussion for learning across the curriculum.* Norwood, MA: Christopher-Gordon.

Kahle, J. B. (2007). Systemic reform: Research, vision, and politics. In S. K. Abell and N. G. Lederman (Eds.), *Handbook of research on science education* (pp. 911–941). Mahwah, NJ: Lawrence Erlbaum.

Kennedy, M. M. (1998). The relevance of content in in-service teacher education. Paper presented at the annual meeting of the American Educational Research Association, San Diego, CA.

Keys, C. W. and Bryan, L. A. (2000). Co-constructing inquiry-based science with teachers: Essential research for lasting reform. *Journal of Research in Science Teaching,* 38(6), 631–645.

Khourey-Bowers, C., and Simonis, D. G. (2004). Longitudinal study of middle grades chemistry professional development: Enhancement of personal science teaching self-efficacy and outcome expectancy. *Journal of Science Teacher Education,* 15(3), 175–195.

Killion, J. (1998). Scaling the elusive summit. *Journal of Staff Development,* 19(4), 12–16.

Knight, P. (2002). A systemic approach to professional development: Learning as practice. *Teaching and Teacher Education,* 18, 229–241.

Kozol, J. (2005). *The shame of the nation: The restoration of apartheid schooling in America.* Pittsburgh: Three Rivers Press.

Kulik, J. A., and Kulik, C. L. C. (1984). Effects of accelerated instruction on students. *Review of Educational Research,* 54(3), 409–425.

Kutal, C., Rich, F., Hessinger, S. A., and Miller, H. R. (2009). Engaging higher education faculty in K-16 stem education reform. In J. S. Kettlewell and R. J. Henry (Eds.), *Increasing the competitive edge in math and science* (pp. 121–134). Lanham, MD: Rowman & Littlefield Education.

Lawrence Hall of Science, University of California, Berkeley. (2003). *Full option science system (FOSS)* (Series). Hudson, NH: Delta Education.

Leavitt, H. J., and Bahrami, H. (1988). *Managerial psychology: Managing behavior in organizations.* Chicago: University of Chicago Press.

Lee, C. D., and Spratley A. (2010). *Reading in the disciplines: The challenges of adolescent literacy.* New York, NY: Carnegie Corporation of New York.

Lee, O. and Lukyx, A. (2006). *Science education and student diversity: Synthesis and research agenda.* New York: Cambridge University Press.

Levine, M. (2002). Why invest in professional development schools? *Educational Leadership,* 59(6), 65–68.

Lewin, K. (1947). Frontiers in group dynamics: Concept, method, and reality in social science, social equilibria, and social change. *Human Relations,* 1(1), 5–41.

Li, J., and Hambrick, D. C. (2005). Factional groups: A new vantage on demographic fault lines, conflict, and disintegration in work teams. *Academy of Management Journal,* 48, 794–813.

Lindamer, L. A., Lebowitz, B. D., Hough, R. L., Garcia, P., Aquirre, A., Halpain, M. C., Depp, C., and Jeste, D. V. (2008). Improving care for older persons with schizophrenia through an academic-community partnership. *Psychiatric Services,* 59, 236–239.

Lipman, P. (2004). *High Stakes Education: Inequity, Globalization, and Urban School Reform.* New York, NY: Taylor and Francis Books, Inc.

Little, J. W. (1982). Norms of collegiality and experimentation: Workplace conditions of school success. *American Educational Research Journal,* 19, 240–325.

Lortie, D. C. (1977). *Schoolteacher: A sociological study.* Chicago: University of Chicago Press.

Loucks-Horsley, S., and Matsumoto, C. (1999). Research on professional development for teachers of mathematics and science: The state of the scene. *School Science and Mathematics,* 99(5), 258–271.

Lynch, S., Kuipers, J., Pyke, C., and Szesze, M. (2005). Examining the effects of a highly rated science curriculum unit on diverse students: Results from a planning grant. *Journal of Research in Science Teaching,* 42(8), 912–946.

Mai, R. (2004). Leadership for school improvement: Cues from organizational learning and *renewal* efforts. *The Education Forum,* 68, 211–221.

Mai, R. and Akerson, A. (2003). *The leader as communicator: Strategies and tactics to build loyalty, focus effort and spark creativity.* New York: AMACOM.

Manning, M. L., Lucking, R., and MacDonald, R. H. (1995). What works in urban middle schools. *Childhood Education,* 71, 221–224.

Marton, F. (1981). Phenomenography—describing conceptions of the world around us. *Instructional Science,* 10, 177–200.

———. (1992). Phenomenography and "the art of teaching all things to all men." *Qualitative Studies in Education,* 5, 253–267.

———. (1994). Phenomenography. In T. Husen and T. N. Postlethwaite (Eds.), *The international encyclopedia of education* (2nd ed., Vol. 8, pp. 4424–4429). Oxford, UK: Pergamon.

———. (1996). Phenomenography—a research approach to investigating different understandings of reality. *Journal of Thought,* 21, 28–49.

Matthews, C. M. (2007). Science, engineering and mathematics education: Status and issues. Congressional research service report for Congress. Order Code 98-871 STM.

McBee, R. H., and Moss, J. (2002). PDS partnerships come of age. *Educational Leadership,* 59(6), 61–64.

McConachie, S. M., Petrosky, A. R., Resnick, L. B. (2009). *Content matters: A disciplinary approach to improving student learning.* San Francisco, CA: Jossey-Bass Education.

McEneaney, E. H. (2003). The worldwide cachet of scientific literacy. *Comparative Education Review,* 47(2), 217–237.

Meier, D. (2002). *The power of their ideas.* Boston: Beacon Press.

Moje, E. B. (2008). Responsive literacy teaching in secondary schools. In M. W. Conley, J. R. Friedhoff, M. B. Sherry, and S. F. Tuckey (Eds.), *Meeting the challenge of adolescent literacy: Research we have, research we need* (pp. 58–81). New York: Guilford Press.

Murawski, W. W. and Swanson, H. L. (2001). A meta-analysis of co-teaching research: Where is the data? *Remedial and Special Education*, 22(5), 258–267.

Nadler, D. A. (2004). Building better boards. *Harvard Business Review*, 82(5), 102–111.

National Center for Education Statistics. (2009). *Achievement gaps: How Black and White students in public schools perform in mathematics and reading on the National Assessment of Educational Progress.* Washington, DC: Institute of Education Sciences.

National Commission on Mathematics and Science Teaching for the 21st Century. (2000). *Before it's too late: A report to the nation.* Washington, DC: Author.

National Comprehensive Center for Teacher Quality. (2006). *Key issue: Recruiting quality mathematics, science, and special education teachers for urban schools.* Washington, DC: Learning Point Associates.

National Council of Teachers of Mathematics. (2000). *Principles and standards for school mathematics.* Reston, VA: Author.

National Research Council. (1996). *National science education standards.* Washington, DC: National Academy Press.

———. (2000). Committee on Developments in the Science of Learning. In J. D. Bransford, A. L. Brown, and R. R. Cocking (Eds.), *How people learn: Brain, mind, experience, and school—Expanded edition.* Commission on Behavioral and Social Sciences and Education. Washington, DC: The National Academies Press.

———. (2007). Committee on Science Learning, kindergarten through eighth grade. In R. A. Duschl, H. A. Schweingruber, and A. W. Shouse (Eds.), *Taking science to school: Learning and teaching science in grades K–8.* Board on Science Education, Center for Education, Division of Behavioral and Social Sciences and Education. Washington, DC: The National Academies Press.

National Science Board. (2006). *America's pressing challenge—Building a stronger foundation.* Arlington, VA: National Science Foundation.

———. (2007). *A national action plan for addressing the critical needs of the U.S. Science, Technology, Engineering, and Mathematics System.* (Rep. No. NSB-07-114), Washington, DC: National Science Foundation.

———. (2009). Actions to improve science, technology, engineering, and mathematics (STEM) education for all American students. Retrieved from http://www.nsf.gov/nsb/publications/2009/01_10_stem_rec_obama.pdf

———. (2010). *Science and Engineering Indicators 2010.* Arlington, VA: National Science Foundation, NSB 10-01.

National Science Foundation. (2010). *FY 2011 budget request to congress.* Retrieved from http://www.nsf.gov/about/budget/fy2011/toc.jsp?org=NSF.

National Science Resources Center. (2009). *Science and technology concepts (STC) program* (Series). Burlington, NC: Carolina Biological Supply.

Nicholson, G. J., and Kiel, G. C. (2004). A framework for diagnosing board effectiveness. *Corporate Governance,* 12, 442–460.

Norman, O., Ault, C. R., Bentz, B.,and Meskimen, L. (2001). The Black–White "achievement gap" as a perennial challenge of urban science education: A sociocultural and historical overview with implications for research and practice. *Journal of Research in Science Teaching,* 38, 1101–1114.

Obama for America. (2009). *The Obama-Biden plan.* Retrieved from http://www.barackobama.com/pdf/issues/PreK-12EducationFactSheet.pdf

O'Brien, D., Stewart, R., and Moje, E. B. (1995). Why content literacy is difficult to infuse into the secondary school: Complexities of curriculum, pedagogy, and school culture. *Reading Research Quarterly,* 30(3), 442–463.

Ogle, D., Klem, R., and McBride, B. (2007). *Building literacy in social studies: Strategies for improving comprehension and critical thinking.* Alexandria, VA: Association for Supervision and Curriculum Development.

Ohio Department of Education. (2007). *Standards for Ohio educators.* Columbus, OH.

Ohio Department of Job and Family Services. (2006). Career opportunities: Occupational trends 2006–2016. Retrieved from http://lmi.state.oh.us/proj/projections/ohio/OccupationalTrends.pdf

Olszewski-Kubilius, P. (2003). Do we fit children to fit gifted programs, or do we change gifted programs to fit gifted children? *Journal for the Education of the Gifted,* 26(4), 304–313.

Partnership for 21st Century Skills. (2004). Retrieved from http://www.21stcenturyskills.org

Patton, M. Q. (2002). *Qualitative evaluation and research methods* (3rd ed.). Thousand Oaks, CA: Sage.

Pellino, K. M. (2007). *The effects of poverty on teaching and learning.* Teachnology, Inc. Retrieved from http://www.teach-nology.com/tutorials/teaching/poverty/print.htm

Pine, G. (2000). Making a difference: A professional development school's impact on student learning. Paper presented at the annual meeting of the American Educational Research Association, New Orleans, LA.

Popham, W. J. (2004). *America's "failing" schools: How parents and teachers can cope with No Child Left Behind.* New York: Routledge Falmer.

Proctor, T. (1999). *Total scores for students on Texas Assessment of Academic Skills 1993–1999.* Waco, TX: Baylor University.

Prybil, L. (2006). Basic characteristics of effective boards. *Trustee,* 59(3), 20–23.

Pye, A., and Pettigrew, A. (2005). Studying board context, process and dynamics: Some challenges for the future. *British Journal of Management,* 16(1), 27–38.

Revak, M., and Kuerbis, P. J. (2008, January). The link from professional development to K-6 student achievement in science, math, and literacy. Paper presented at the Annual International Meeting of the Association for Science Teacher Education, St. Louis, MO.

Rimm, S. B. (1986). *Underachievement syndrome: Causes and cures.* Watertown, WI: Apple Publishing.

Robbins, S. (2003). *Organizational Behavior.* 10th ed. Upper Saddle River, NJ: Prentice Hall.

Rogers, K. B. (1991). *The relationship of grouping practices to the education of the gifted and talented learner* (executive summary). Storrs, CT: National Research Center on the Gifted and Talented.

Rose, D. H. and Meyer, A. (2002). *Teaching every student in the digital age: Universal Design for Learning.* Alexandria, VA: ASCD.

Rowley, S. J. and Moore, J. A. (2002). Racial identity in context for the gifted African American student. *Roeper Review,* 24, 63–67.

Ruby, A. (2006). Improving science achievement at high-poverty urban middle schools. *Science Education,* 90(6), 1005–1027.

Rutherford, J. F. (2005). The 2005 Paul F-Brandwein lecture: Is our past our future? Thoughts on the next 50 years of science education reform in the light of judgments on the past 50 years. *Journal of Science Education Technology,* 14, 367–386.

Sanders, W. L. and Wright, S. P. (2008). *Methodological concerns about the education value-added assessment system.* Knoxville: University of Tennessee Value Added Research and Assessment Center.

Schmid, H. (2006). Leadership styles and leadership change in human and community service organizations. *Nonprofit Management & Leadership,* 7, 179–194.

Schoenfeld, A. (2009). Working with schools: The story of a mathematics education collaboration. *American Mathematical Monthly,* 116(3), 197–217.

Scribner, J. P., Sawyer, R. K., Watson, S. T., and Myers, V. L. (2007). Teacher teams and distributed leadership: A study of group discourse and collaboration. *Educational Administration Quarterly,* 43(1), 67–100.

Seidman, I. (1998). *Interviewing as qualitative research: A guide for researchers in education and the social sciences.* New York, NY: Teachers College Press.

Settlage, J. and Meadows, L. (2002). Standards based reform and its unintended consequences: Implications for science education within America's urban schools. *Journal of Research in Science Teaching,* 39(2), 114–127.

Shanahan, T., and Shanahan, C. (2008). Teaching disciplinary literacy to adolescents: Rethinking content-area literacy. *Harvard Educational Review,* 78(1), 40–59.

Shelley, M. C., II. (2009). Speaking truth to power with powerful results: Impacting public awareness and public policy. In M. C. Shelley II, L. D. Yore, and B. Hand (Eds.), *Quality research in literacy and science education: International perspectives and gold standards* (pp. 443–466). Dordrecht, The Netherlands: Springer.

Shirley, D. (1997). *Community organizing for urban school reform.* Austin, TX: University of Texas Press.

———. (2009). Community organizing and education change: A reconnaissance. *Journal of Educational Change,* 10, 229–237.

Shymansky, J. A., Annetta, L. A., Yore, L. D., Wang, T.-L., and Everett, S. A. (2010). The impact of a multi-year systemic reform effort on elementary school students' achievement in science (Manuscript submitted for publication).

Shymansky, J. A., Wang, T.-L., Annetta, L. A., Yore, L. D., and Everett, S. A. (2010a). How much professional development is needed to effect positive gains in K-6 student achievement on high stakes science tests? *International Journal of Science and Mathematics Education.* Advance online publication. doi: 10.1007/s10763-010-9265-9

———. (2010b). The impact of a multi-year, multi-school district K-6 professional development program designed to integrate science inquiry and language arts on students' high stakes test scores (Manuscript submitted for publication).

Shymansky, J. A., Yore, L. D., and Anderson, J. O. (2004). Impact of a school district's science reform effort on the achievement and attitudes of third- and fourth-grade students. *Journal of Research in Science Teaching,* 41(8), 771–790.

Shymansky, J. A., Yore, L. D., and Hand, B. (2000). Empowering families in hands-on science programs. *School Science and Mathematics,* 100(1), 48–56.

Shymansky, J. A., Yore, L. D., Treagust, D. F., Thiele, R. B., Harrison, A., Waldrip, B. G., Venville, G. (1997). Examining the construction process: A study of changes in level 10 students' understanding of classical mechanics. *Journal of Research in Science Teaching,* 34(6), 571–593.

Smagorinsky, P. (2001). If meaning is constructed, what is it made from? Toward a cultural theory of reading. *Review of Educational Research,* 71(1), 133–169.

Smith, F. (1988). *Joining the literacy club: Further essays into education.* Portsmouth, NH: Heinemann.

Smith, P., and Tompkins, G. (1988). Structured notetaking: A new strategy for content area readers. *Journal of Reading,* 32, 46–53.

Southern, W. T., Jones, E. D., and Stanley, J. C. (1993). Acceleration and enrichment: The *context* and development of program options. In K. A. Heller, F. J. Mönks, and A. H. Passow (Eds.), *International handbook of research and development of giftedness and talent* (pp. 387–405). New York: Pergamon.

SRI International. (2009). Making an impact: Assessing the benefits of Ohio's investment in technology-based economic development programs. Retrieved from http://www.development.ohio.gov/ohiothirdfrontier/Documents/Recent Publications/OH_Impact_Rep_SRI_FINAL.pdf

Stahl Ladbury, J. L., Hall, B. S., and Benz, R. A. (2010). Transient students: Addressing relocation to a new high school through group counseling. *Ideas and research you can use: VISTAS 2010.* Retrieved from http://counselingoutfitters .com/vistas/vistas10/Article_58.pdf

Stajkovic, A. D., Lee, D., and Nyberg, A. J. (2009). Collective efficacy, group potency and group performance: Meta-analysis of their relationships and test of a mediation model. *Journal of Applied Psychology,* 94(3), 814–828

Stepanek, J. (1999). *The inclusive classroom. Meeting the needs of gifted students: Differentiating mathematics and science instruction.* Portland, Oregon: Northwest Regional Educational Laboratory.

Stevens, M. J., and Campion, M. A. (1994). The knowledge, skill and ability requirements for teamwork: Implications for human resource management. *Journal of Management,* 20, 503–530.

Stewart, G. L., Fulmer, I. S., and Barrick, M. R. (2005). An exploration of member roles as a multilevel linking mechanism for individual traits and team outcomes. *Personnel Psychology,* 58, 343–365.

Stiles, P., and Taylor, B. (2001). *Boards at work: How directors view their roles and responsibilities.* Oxford, UK: Oxford University Press.

Stoddart, T., Pinal, A., Latzke, M., and Canaday, D. (2002). Integrating inquiry science and language development for English Language Learners. *Journal of Research in Science Teaching,* 39(8), 664–687.

Stufflebeam, D. L., Foley, W. L., Gephart, W. J., Guba, E. G., Hammond, R. L., Marriman, H. O., and Provus, M. M. (1971). *Educational evaluation & decision making: Phi delta kappa national study committee on evaluation.* Itasca, IL. Peacock Publishers.

Sundaramurthy, C., and Lewis, M. (2003). Control and collaboration: Paradoxes of governance. *Academy of Management Review,* 28, 397–415.

Supovitz, J. A., and Turner, H. M. (2000). The effects of professional development on science teaching practices and classroom culture. *Journal of Research in Science Teaching,* 37(9), 963–980.

Swackhamer, L. E., Koellner, K. A., Basile, C. G., and Kimbrough, D. (2009). Increasing the self-efficacy of in-service teachers through content knowledge. *Teacher Education Quarterly,* 36(2), 63–78.

Tabachnick, B. G., and Fidell, L. S. (2007). *Using multivariate statistics* (5th ed.). Boston, MA: Allyn & Bacon.

Tasa, K., Taggar, S., and Seijts, G. H. (2007). The development of collective efficacy in teams: A multilevel and longitudinal perspective. *Journal of Applied Psychology,* 92(1), 17–27.

Taylor, P. M. (2004). Encouraging professional growth and mathematics reform through collegial interaction. In R. N. Rubenstein and G. W. Bright (Eds.), *Perspectives on the teaching of mathematics* (pp. 219–228). Reston, VA: NCTM.

Teitel, L. (1993). The state role of jump-starting school/university collaboration: A case study. *Educational Policy,* 7(1), 74–95.

The Higher Education Opportunity Act (Public Law 110-315). Retrieved December 9, 2009, from http://www.ed.gov/policy/ highered/leg/hea08 /index.html

The No Child Left Behind Act of 2001. (2001). Publication L. No. 107-110.

The White House. (2009). President Obama launches "educate to innovate" campaign for excellence in science, technology, engineering & math (STEM) education. The White House office of the press secretary. Retrieved from http://www.whitehouse.gov/the-press-office/president-obama-launches-educate-innovate-campaign-excellence-science-technology-en

———. (2010). President Obama expands "educate to innovate" campaign for excellence in science, technology, engineering, and mathematics (STEM) education. The White House office of the press secretary. Retrieved from http://www.whitehouse.gov/the-press-office/president-obama-expands-educate-innovate-campaign-excellence-science-technology-eng

Thompson, C. L., and Zeuli, J. S. (1999). The frame and the tapestry: Standards-based reform and professional development. In L. Darling-Hammond and G. Sykes (Eds.), *Teaching as the learning profession: Handbook of policy and practice* (pp. 341–375). San Francisco: Jossey-Bass.

Thompson, S. M., Ransdell, M., and Rousseau, C. (2005). Effective teachers in urban school settings. *Journal of Authentic Learning* 2(1), 22–36.

Triangle Coalition for Science and Technology Education. (2005). *Comparisons of the department of education and national science foundation math and science partnerships.* Retrieved from http://www.trianglecoalition.org/resources.htm

Ullucci, K. (2009). "This has to be family": Humanizing classroom management in urban schools. *Journal of Classroom Interaction,* 44(1), 13–28.

UNESCO. (2008). *UNESCO survey finds under-privileged children also disadvantaged in the classroom.* Montreal, Quebec, Canada: UNESCO Institute for Statistics. Retrieved from http://www.unescobkk.org/fileadmin/user_upload/library/OPI/Documents/UNESCO_in_the_news/0805May28UIS.pdf

U.S. Commission on National Security/21st Century. (2001). Road map for national security: Imperative for change. Washington, DC.

U.S. Department of Education. (1991). *America 2000: An education strategy.* Washington, DC.

———. (2002). *NCLB and other elementary/secondary policy documents.* Retrieved May 15, 2009, from http://www.ed.gov/policy/elsec/guid/states/index.html#nclb

———. (2009). *Race to the top executive summary.* Retrieved from http://www2.ed.gov/programs/racetothetop/index.html

———. (2010). *Mathematics and science partnerships summary of performance period 2007 annual reports.* Retrieved from http://www.ed-msp.net/index.php?option=com_content&view=article&id=5&Itemid=2

vanDriel, J. H., Beijaard, D., and Verloop, N. (2001). Professional development and reform in science education: The role of teachers' practical knowledge. *Journal of Research in Science Teaching,* 38(2), 137–158.

Wallis, C. (2006). How to bring our schools out of the 20th century. *TIME Magazine.* December 2006.

Weinburgh, M. E. and Steele, D. (2000). The modified attitudes toward science inventory: Developing an instrument to be used with fifth grade urban students. *Journal of Women and Minorities in Science and Engineering,* 6, 87–94.

Wiggins, G. and McTighe, J. (2005). *Understanding by design.* New Jersey: Prentice Hall.

Willging, C. E., Tremaine, L., Hough, R. L., Reichman, J. S., Adelsheim, S., Meador, K., and Downes, E. A. (2007). "We never used to do things this way": Behavioral health care reform in New Mexico. *Psychiatric Services,* 58(12), 1529–1531.

Willis, S. (2002). Creating a knowledge base for teaching: A conversation with James Stigler. *Educational Leadership,* 59(6), 6–11.

Wilson, S. M., and Berne, J. (1999). Knowledge: An examination of research on contemporary professional development. *Review of Research in Education,* 24, 173–209.

Wolcott, H. F. (1994). Transforming qualitative data: Description, analysis, and interpretation. Thousand Oaks, CA: Sage Publications.

Wright, J. C. and Wright, C. S. (2000). A commentary on the profound changes envisioned by the national science standards. *Teachers College Record,* 100(1), 122–143.

Yin, R. K. (2003). *Case study research: Design and methods* (3rd ed.). Thousand Oaks, CA: Sage Publications.

Yore, L. D. (2009). Science literacy for all: More than a logo or rally flag! In K. C. D. Tan and Y. J. Lee (Eds.), *International Science Education Conference 2009* (pp. 2393–2427). Singapore: National Institute of Education. Retrieved from http://www.nsse.nie.edu.sg/isec2009/downloads/

Yore, L. D., Anderson, J. O., and Shymansky, J. A. (2005). Sensing the impact of elementary school science reform: A study of stakeholder perceptions of implementation, constructivist strategies, and school-home collaboration. *Journal of Science Teacher Education,* 16(1), 65–88.

Yore, L. D., Henriques, L., Crawford, B., Smith, L., Gomez-Zwiep, S., and Tillotson, J. (2007). Selecting and using inquiry approaches to teach science: The influence of context in elementary, middle, and secondary schools. In E. Abrams, S. A. Southerland, and P. Silva (Eds.), *Inquiry in the classroom: Realities and opportunities* (pp. 39–87). Greenwich, CT: Information Age.

Yore, L. D., Pimm, D., and Tuan, H.-L. (2007). The literacy component of mathematical and scientific literacy (special issue). *International Journal of Science and Mathematics Education,* 5(4), 559–589.

Zmuda, A., Kuklis, R., and Kline, E. (2004). *Transforming schools: Creating a culture of continuous improvement.* Alexandria, Virginia: Association for Supervision and Curriculum Development.

Contributors

Len Anetta is an associate professor of science education at George Mason University. Dr. Annetta's research has focused on distance learning and the effect of instructional technology on science learning of teachers and students in underserved populations. Dr. Annetta has twice been awarded the National Technology Leadership Initiative Fellowship in Science Education and Technology from the Association of Science Teacher Education and the Society for Information Technology and Teacher Education.

James Basham is an assistant professor of special education at the University of Kansas. He earned his doctorate at the University of Illinois Urbana-Champaign. Dr. Basham's research is focused on student learning in modern learning environments chiefly related to the application of Universal Design for Learning (UDL). He serves on the editorial board for the *Journal of Special Education Technology* and is a reviewer for the *Journal of Educational Computing Research*. Dr. Basham was the recent co-guest editor for the JSET topical issue on STEM education for individuals with diverse learning needs.

Virginia L. J. Bolshakova is a doctoral student at Utah State University in ecology. Bolshakova's master's degree was received from the University of Toledo in curriculum and instruction with a specialization in science education. Her bachelor's degree is from Utah State University in secondary biology education. Bolshakova's research is focused in the areas of self-efficacy in science education.

Lawrence J. Coleman is the Herb Professor of Gifted Education in the Judith Herb College of Education, Health Science and Human Service at the University of Toledo. He is a special education teacher who became a professor and has spent most of his career studying and writing about children who are gifted and talented.

Charlene M. Czerniak is a professor at the University of Toledo in the Department of Curriculum and Instruction. She received her PhD in science education from the Ohio State University. Professor Czerniak has authored and coauthored over 50 articles. Professor Czerniak is coauthor of a textbook published by Lawrence Erlbaum and Associates on project-based science teaching. She also has 5 chapters in books and illustrated 12 children's science education books. For five years, she served as editor of the *Journal of Science Teacher Education*, the professional journal of the Association for Science Teacher Education. Charlene Czerniak was the president of the School Science and Mathematics Association for two years, and she served as the president of the National Association for Research in Science Teaching (NARST) from 2008–2009. In 2010, she received the George Mallinson Distinguished Service Award from the School Science and Mathematics Association (SSMA), which is the highest award given by SSMA.

Kadir Demir is an assistant professor of science education in the Department of Middle and Secondary Education and Instructional Technology Department at Georgia State University. Dr. Demir holds a bachelor's degree in biology teaching, from Gazi University-Turkey, and two master's degrees, one in science education and one in educational technology, and a PhD in science education from University of Missouri-Columbia. He teaches undergraduate and graduate classes in science education. His current research foci include reform-based practices of college science faculty and pre-/in-service secondary science teachers and science education in urban settings.

Susan Everett is an associate professor of science education in the School of Education at the University of Michigan-Dearborn. She is the coordinator of the master of science in science education and leads the Inquiry Institute at UM-D. Dr. Everett is a contributing editor to *Science Scope* with the featured column "Everyday Engineering" and is a member of the editorial board for *International Journal of Science and Mathematics Education*. Her research interests focus on inquiry-based science teaching and learning.

Shelly S. Harkness is an associate professor, secondary education mathematics, at the University of Cincinnati and has taught both mathematics and methods courses. She was a classroom teacher for 12 years prior to earning her PhD at Indiana University. Her research interests include ethnomathematics; mathematics and social justice; and teachers' "believing" and listening in mathematics classrooms. Dr. Harkness is coeditor of the *School Science and Mathematics* journal.

Maya Israel is an assistant professor of special education at the University of Cincinnati. She teaches courses in the School of Education related to content-area literacy, special education, and secondary science methods. Her primary areas of specialization include technology integration into teachers' instructional practices, Universal Design for Learning (UDL), and e-mentoring of early career educators. Her work includes a U.S. Department of Education grant that explores how teachers and content developers can provide accessible UDL-based STEM instructional content to students with diverse learning needs and an Ohio Board of Regents grant that explores how school districts can provide STEM learning opportunities to their students.

Carla C. Johnson is an associate professor of science education, as well as coordinator of the Curriculum and Instruction and Middle Childhood Education Programs at the University of Cincinnati. Dr. Johnson's research is in the areas of effective science instruction and professional development, STEM education policy, practice, and reform, as well as culturally relevant science pedagogy. Dr. Johnson was the 2006 recipient of the Outstanding Early Career Scholar Award from the School Science and Mathematics Association and currently serves as editor of the *School Science and Mathematics* journal and on the editorial boards of many journals in her field.

Holly Johnson is an associate professor of literacy and director of the School of Education at the University of Cincinnati. Johnson received her PhD in language, reading, and culture at the University of Arizona. Her work in higher education focuses on adolescents' literacy, inquiry learning across content disciplines, and disciplinary literacy. Prior to entering higher education, Dr. Johnson taught middle school language arts in Arizona and Kentucky. She was a teacher participant in the Human Biology Project while in Kentucky.

Catherine Koehler is an assistant professor of science education at the Illinois Institute of Technology in Chicago Illinois. Dr. Koehler's research explores systemic reform efforts in the areas of nature of science and scientific inquiry, nature of engineering, magnet school reform, STEM education policy and practice, and urban education.

Andrea R. Milner is an assistant professor at Adrian College in the Teacher Education Department. Dr. Milner received her PhD in curriculum and instruction from the University of Toledo. Dr. Milner's line of inquiry is investigating the effects constructivist classroom contextual factors have on student motivation and learning strategy use.

Catherine Pullin Lane is a mathematics instructor in the Science, Mathematics and Engineering Division at Clermont College. She has taught both mathematics and mathematics education courses. She is also a doctoral student in curriculum and instruction at the University of Cincinnati. Her research interests include how students develop mathematical thinking skills and their interactions with the concept of proof.

James A. Shymansky is an E. Desmond Lee Professor of Science Education at the University of Missouri-St. Louis in the Institute for Mathematics, Science and Educational Learning Technology. He has authored or coauthored more than 25 books, chapters, and monographs, an elementary science series, and more than 80 journal articles. He is an active member of the National Association for Research in Science Teacher where he has served as president and editor of the *Journal of Research for Science Teaching.* He is currently serving as senior editor of the *International Journal of Science and Mathematics Education* and as director of the "Just ASK" project, a DR K-12 project funded by the National Science Foundation.

Toni A. Sondergeld is an assistant professor of assessment, research, and statistics at Bowling Green State University in the College of Education and Human Development's School of Leadership and Policy Studies. She started her teaching career as a junior high science educator. Her research interests include urban school reform and classroom assessment.

W. Thomas Southern is a professor of special education in the Department of Educational Psychology at Miami University in Oxford, Ohio. He has spent most of his career looking at the identification of and programing for special populations of gifted children.

Lacey Strickler is a doctoral student at the University of Toledo in the Curriculum and Instruction Department focusing her work in science education. Previously she completed a bachelors degree in biology and a masters in molecular and cellular biology. Strickler has experience working within the informal science education field, and her research interests include science learning in informal settings and undergraduate biology education.

Camille Sutton-Brown is a doctoral candidate in the Department of Educational Policy Studies at Georgia State University. Camille Sutton-Brown holds a bachelor's degree in family and social relations from the University of Guelph in Ontario, Canada and a master's degree in special education from Georgia State University. Her research interests include sustainable community development, women's empowerment, and the use of visual methodologies to improve instructional practices.

Patricia Watson is an associate professor of reading in the College of Professional Education at Texas Woman's University in Denton, TX. A former middle school language arts and technology teacher, she currently works with preservice and in-service teachers in the Dallas-Fort Worth Metroplex.

Larry D. Yore is a University Distinguished Professor in the Faculty of Education at the University of Victoria, Victoria, British Columbia, Canada. In 47 years of teaching and researching at elementary, secondary, and tertiary levels, he has been engaged in developing provincial science curricula, national science frameworks, national K-12 assessment projects, and several federally funded teacher enhancement, reform, and research and development projects in Canada and the United States. His recent research focuses on the constructive, communicative, and persuasive roles of language in doing and learning science. He has published numerous journal articles, edited books, and chapters; coauthored elementary science textbooks; edited special issues related to applications of language arts in science education, PISA results, and mathematical and scientific literacies; and presented numerous lectures.

Index